TRENDS IN GEOMATICS - AN EARTH SCIENCE PERSPECTIVE

Edited by **Rifaat Abdalla**

Trends in Geomatics - An Earth Science Perspective
http://dx.doi.org/10.5772/intechopen.75730
Edited by Rifaat Abdalla

Contributors

M. Atif Butt, Syed Amer Mahmood, Jahanzeb Qureshi, Muhammad Kashif Nazir, Amer Masood, Javed Sami, Hou-Pu Li, Shao-Feng Bian, Oscar Frausto-Martinez, Orlando Colin - Olivares, Norma Angélica Zapi - Salazar, Rifaat Abdalla

Notice

Statements and opinions expressed in the chapters are these of the individual contributors and not necessarily those of the editors or publisher. No responsibility is accepted for the accuracy of information contained in the published chapters. The publisher assumes no responsibility for any damage or injury to persons or property arising out of the use of any materials, instructions, methods or ideas contained in the book.

First published in London, United Kingdom, 2019 by IntechOpen
IntechOpen is the global imprint of INTECHOPEN LIMITED, registered in England and Wales, registration number: 11086078, The Shard, 25th floor, 32 London Bridge Street
London, SE19SG – United Kingdom
Printed in Croatia

British Library Cataloguing-in-Publication Data
A catalogue record for this book is available from the British Library

Additional hard copies can be obtained from orders@intechopen.com

Trends in Geomatics - An Earth Science Perspective, Edited by Rifaat Abdalla
p. cm.
Print ISBN 978-1-78985-435-0
Online ISBN 978-1-78985-436-7

We are IntechOpen,
the world's leading publisher of
Open Access books
Built by scientists, for scientists

4,000+
Open access books available

116,000+
International authors and editors

120M+
Downloads

Our authors are among the

151
Countries delivered to

Top 1%
most cited scientists

12.2%
Contributors from top 500 universities

Interested in publishing with us?
Contact book.department@intechopen.com

Numbers displayed above are based on latest data collected.
For more information visit www.intechopen.com

Meet the editor

Dr. Rifaat Abdalla is an Associate Professor in the Department of Earth Sciences, College of Science, Sultan Qaboos University. He specialized in GeoInformatics with a focus on WebGIS applications and remote sensing modeling. Dr. Rifaat Abdalla is a Certified GIS Professional (GISP) and Professional Geoscientist (P.Geo.) from Ontario, Canada. He was with the King Abdulaziz University, Jeddah for five years prior to joining Sultan Qaboos University in 2017. Dr. Abdalla worked in the oil industry in Qatar and in the UK and has taught at York University and Ryerson University in Canada. His research interests include hydrography and marine applications for GIS; modeling and simulation; mobile handheld GIS visualization; and disaster management and emergency response mapping. Dr. Abdalla is a recipient of several prestigious international awards, including the American Society for Photogrammetry and Remote Sensing (ASPRS) PE&RS Best Scientific Paper by ESRI.

Contents

Preface

The need for geomatics technologies and data products is on the rise. This book provides brief coverage on the wide spectrum of geomatics and geomatics data products and procedures. The importance of geomatics technologies arises from the evaluation of advanced tools and an approach to providing geospatial knowledge to a variety of applications including decision support. The data collected from different geomatics tools is undergoing growing demand, and it requires maintenance and updating. Regardless of the data utilization sector or the exploitation process, data derived from geomatics approaches, such as remote sensing and land surveying provide a wealth of knowledge and thus require wide dissemination for decision support as well as for public access.

The scope, application and level of adoption of geomatics technologies provides information about the domain-based inventory of spatial data, products, services and systems available for a specific application. This can be challenged by the need for identification of the standards and procedures in accordance with the future directions in the domain of interest with emphasis on accuracy, precision and update. Regular assessment of the level of integration between different geomatics sectors is essential in assessment of the standard procedures and implementation options.

Knowledge-based and modeling approaches help improve the interdisciplinary capabilities of geomatics technologies. Various levels of knowledge can be acquired through geomatics technologies. For instance, remotely sensed imagery and 3D models content management, discussion forms, interactive map displays, user comments and feedback and decision support, events scheduling and spatially enabled calendar of events that provides sharing of location information on social networks, and location-based search engine functionalities, and electronic newsletter sharing, as well as backup and restore functions.

It is not only data issues that are associated with effective implementation of geomatics technologies. It can also be the system architecture, including content management, GIS, data, web portal and integration components. Communications, hardware, and hosting. System security plan including system configuration, Web site map design, including number of map views, layers and functionality. Database design, including shared design aspects, and specific domain-related tables. A service directory to support initial implementation and serve as framework for future development of the system.

We hope that this book will provide a benefit to the professionals and researchers in geomatics, as it highlight some important trends that are essential to earth scientists and those who are interested in a more applications-based coverage from interdisciplinary approaches.

Rifaat Abdalla, Ph.D., P.Geo., GISP.
Department of Earth Sciences
Sultan Qaboos University, Oman

Introduction Chapter: From Land Surveying to Geomatics - Multidisciplinary Technological Trends

Rifaat Abdalla

Additional information is available at the end of the chapter

http://dx.doi.org/10.5772/intechopen.82424

1. Growing technological trends in geomatics

Substantial work has been done by the geomatics community to evaluate and develop mechanisms and interrelationships that are required for integrating data, procedures, and protocols in an effective way. Targeting higher levels of integration and coordination with domains is the link that is found to be efficient in connecting geomatics communities, remote sensing, GIS, and global navigation systems. It attempts to achieve the desired level of effectiveness by providing low-cost, highly efficient, highly accurate systems. This integration can help with the ability of freely exchanging all kinds of geospatial data and information, through accessing various data sources and systems over networks. This includes the ability of processing large volumes of data and analyzing it through utilization of effective hardware and software. This can help with achieving a combined decision-making approach which incorporates all user classes.

The growing trends in geomatics technology are of special importance, because they bring many challenges and opportunities to the user community. The user community can be application users, developers, or technology integrators, in addition to educators. All these communities are contributing to the development of geomatics through providing new approaches and means that leads to progressive advancement in the field, especially in the last few decades. An increasingly important and popular development is in geomatics education and the use of technological means to equip future generations with the foundations to excel in this important domain. Geomatics, hydrography, or applied earth sciences all address the fields of geomatics, whether it be global navigation and positioning technologies, geospatial information technologies, remote sensing technologies, or photogrammetry and field surveying technologies, together with advanced training that supports the moving trends in the important domain. Four major aspects of technology development share the importance of helping wider application and the use of geomatics: (1) environmental monitoring and detecting what changes have occurred, over a specific timeframe, (2) identifying and filling gaps in the user community application cycle, (3) measuring and planning for few advancement in form software and hardware to address the growing user community, and (4) assessing the current level of utilization and adoption of technology. Techniques to utilize geomatics technologies to solve day-to-day issues are on the rise; as such the ease of technological tools are also on the rise, because of increasing versatility in manipulating digital data and increasing computing power. Still, some work needs to be done for the determination of the qualitative accuracy of the different geomatics data collection tools, along with ancillary data acquisition system. The primary objective of this work was to highlight the different levels of ongoing trends in the field of geomatics technology and what type of change has occurred over the recent decades. The main objectives were (1) to provide coverage for some of the important trends in geomatics technology that might influence the user community in all the aspects associated with the themes of geomatics collectively, (2) to highlight the issue of trends in geomatics education, and (3) to perform an evaluation of the role of public participation in geomatics technologies.

2. Data remains central in geomatics

Geomatics aims to secure an automated process which will allow for using different data products, services, and tools across and beyond organizational boundaries. This helps by making data and information available for all the three levels involved in decision-making and acquisition processing, whether at the local level, provincial level, or at the state or country level. Geomatics systems provide integration between hardware, software, and user that take into consideration spatial data storage manipulation, analysis, and visualization. The Internet has played a major role in connecting systems together over a common network protocol, that is, Transmission Control Protocol-Internet Protocol (TCP/IP). This protocol has truly revolutionized the era of information technology; as a result a great deal of interest toward disaster management applications using Internet infrastructure is rapidly evolving. This advancement in Internet technology in addition to the other advances in high-speed broadband Internet-added capabilities has significantly contributed for data interoperability for disaster management applications. Geomatics is a science related to the techniques related to urban data in its digital form, which includes spatial or geographic information systems, including the collection of urban information and work on processing, analysis, and presentation, and the formation of maps and management of urban data. Due to the evolution of information and communication technologies, this has led to the development of geographic information systems and maps and geomatics. Traditional geography has evolved, and Earth observation activities have evolved considerably. Satellites are being used for Earth observation and remote sensing. The use of modern technologies results in a huge amount of data that must be processed and analyzed so that they can be used, and this data is processed in several ways. Geomatics (geos: Earth; matics: informatics) can satisfy such requirements.

Geomatics include the fields of mapping, surveying, remote sensing (LiDAR or HDS scanning), hydrography, photogrammetry, global positioning systems (GPS), and geographic information systems (GIS).

3. Location and user information

In geomatics location is the main factor which is used to integrate a very wide range of data for spatial analysis and visualization. Geometrics engineers apply geometrical principles to spatial information, managing spatial data infrastructures of various types, whether local, regional, or global. The extensive availability and the use of sophisticated technologies, such as global navigation satellite systems (GNSS), remote sensing, and geographical information systems (GIS), increase the precision and productivity of the profession.

Although land surveying uses direct contact with everything measured, remote sensing is a type of measurement that collects data from the air using LiDAR, from the ground with high-definition laser scanning (HDS), or from an automobile using mobile mapping. Although LiDAR can survey high large areas in a short time, it is limited to the acquisition system altitude and swath. On the other hand, HDS equipment mounted on survey tripods is operated similarly to a surveyor's robotic total station. The move from one control point to another can gather up to 50000 points per second, depending on the scanner of the data. HDS has found many applications in different fields from architectural modeling, historical preservation, and civil engineering design to food processing and manufacturing, industrial renovation, and mechanical engineering designs. Photogrammetry combines with LiDAR oftentimes by the high-resolution pictures taken from an aircraft, which is hundreds of feet from the ground. The overlapping of images and their association with aerial panels to control the surface survey result in data processing, correction, and utilization. Hydrography uses a combination of land-based GPS control, ship-mounted GPS receivers, and sounding equipment to map accurately the floor of bodies of water. The use of these geomatics elements or any combination thereof, together under a layer scheme for design purposes or into a geographic information system (GIS), leads to best analysis of our three-dimensional world. Database information are combined together with spatial information about a particular structure or object feature within GIS to provide for analyzing or tracking of features in our environment. Geodesy is a mathematical science that determines the shape and size of the earth and the nature of the earth's gravity. The use of geomatics and its sophistication with the techniques are evolving and continue because of the need to integrate them with modern technologies to be used in several fields, including artificial intelligence, geo-analysis, and geospatial information. The availability of more sensors as a result of their low cost under the Internet, free and open source objects and software, and the availability of high-performance infrastructure all led to the development of geometrics. It is a science that includes many important aspects, so we focus on all geomatics trends.

4. Is there any way forward?

Geomatics approaches and products have been widely used for many applications. This book targets many groups that are of interest to the geomatics user community. This book highlights various trends in the user side, focusing on public participation (GIS); in the geodesy and navigation sides, focusing on some mathematical modeling for geodesy; and in the growing trend of geomatics education, and a focus to provide detailed knowledge for future generation on best available solutions and best practices that utilize geomatics technologies is needed.

Author details

Rifaat Abdalla

Address all correspondence to: rabdalla@squ.edu.om

Department of Earth Sciences, College of Science, Sultan Qaboos University, Oman

Architectural Design and Prototyping of Co-PPGIS: A Groupware-Based Online Synchronous Collaborative PPGIS to Support Municipality Development and Planning Management Workflows

Muhammad A. Butt, Syed Amer Mahmood,
Javed Sami, Jahanzeb Qureshi,
Muhammad Kashif Nazir, Amer Masood,
Khadija Waheed and Aysha Khalid

Additional information is available at the end of the chapter

http://dx.doi.org/10.5772/intechopen.80091

Abstract

Co-PPGIS has a wide variety of applications like municipal planning, emergency response, public health and security, etc. The main focus of this chapter is on the development and design of a Web Collaborative PPGIS (Co-PPGIS) infrastructure. As part of municipality's planning and management services, Co-PPGIS is developed for real-time map sharing application system. Co-PPGIS is an effective and essential online meeting system for supporting group collaborations on geographic information such as maps and imageries, and capturing and sharing of local/domain knowledge in real time. Co-PPGIS permits amalgamation of geospatial data and collaborator's input in the form of geo-referenced notations. It incorporates coherent components as map sharing, real-time chat, video conferencing, geo-referenced textual and graphical notations. The study aims to focus on public participation and geo-collaboration facilitated with information sharing, interactive geo-conferencing, real-time map, and data sharing with tools to draw features or add annotation to the map while discussions, uploading documents, and live communication. Co-PPGIS provides an efficient and reliable platform that will significantly reduce the time to acquire, process, and analyze data. The significance of this study is to contribute to existing public participation practices, to municipal planning, to decision-making, or to geographic information science.

Keywords: PPGIS, GIS, GSC, participation, feedback, WWW, CSCW

1. Introduction

1.1. Study context

In recent years, providing public role in decision-making regarding spatial problems has developed an ease for geographic information technology adept in supporting collaborative spatial decision-making. According to Densham et al. [1], it has been stated that Geographical Information System is the technology to sustain PPGIS, but expert methods are needed to reinforce spatial decision-making in a collective way. However, now, Geographical Information System and PPGIS are not prototyped to assist multi-users associations but many approaches may require group-based involvement for decision-making. The idea of collaborative geographical information system, computer-supported cooperative work (CSCW), and collaborative decision support systems (DSS) were proposed as information technology to provide understanding about spatial complications and provide computer-based spatial decision-making [2].

Multi-user collaboration is playing its role in many works involving stakeholders from different departments and organizations, in which map making sometimes play a main role for giving visual information for the support to decision-making [3]. Web technology is rapidly expanding range and has made it possible for to take decisions over the Web. Due to demands for Web-based open mapping an Application Programming Interface (API) united with other information systems and CSCW tools have become more important for the support of real-time map sharing output. Accordingly, the development of map-based applications for real-time collaborative is one effective step taken by researchers that are efficiently working in many fields, e.g., emergency system, urban planning projects, municipality management, GIS data production, monitoring of urban sprawl and epidemic spread, and many more that assimilate collaborative role [4–9].

A concurrent approach is made for the support of collaboration among the users [10]; although, little work has been done on developing and designing such Open Source Software (OSS) which is based on online map sharing tools which supports real-time collaboration. By assessing the researchers work and their contributions from the literature review, this study aims to develop an outline about the significance of the execution of irreplaceable and sufficient methods, tools, and techniques to fill the gap in the research. Multi-user synchronous discussions and communications among the people and between the community and stakeholders sometime improve the understanding that show an effective feedback and magnify decision-making [10–14]. This chapter actually shows a customizable framework used for an online system for collaboration with the installation of different Web GIS, OSGIS, OSS-based tools, and open mapping APIs on geographic information to solve the issues that are related to emergency disaster occurrence and municipal planning. Additionally, the study anticipates designing an open mapping API based real-time collaborative synchronous infrastructure with the option of installing local data for improving the involvement of during debate. Some of this research prototype elements based on this kind of model is still in development procedure and in its starting stage in the house applicable testing.

1.2. Study objectives

The study aims to develop a real-time map sharing mechanism, collaborative PPGIS (Co-PPGIS) and for collaborative assessment the amalgamation of other open source-based groupware solutions on an effective GIS-based meeting platform. The aim of this study was also to assure that: (1) Co-PPGIS model will help to improve or increase involvement of participants and will provide assistance to decision makers in reaching a final decision efficiently; (2) to explain certain facts or observations, i.e., core concepts, design and technology, etc., with an overview of enabling technologies for analyzing and designing a successful real-time map sharing framework; (3) to describe a prototype development based on case scenarios that looks into integrating CSCW principles and open source groupware tools with Web-based GIS. In order to assist municipal planning and development through a better and effective decision-making process the primary research goal is to develop a Web GIS-based contemporary collaborative participatory infrastructure. In order to fulfill the main research's goal, this study will focus on achieving the following objectives:

1. To gain better and effective understanding of the PPGIS' nature, its culture, its limitations, and basic requirements by modeling general as well as high-level participation requirements after proper and complete analysis of municipal Planning and Development (P & D) process workflows and by reviewing the existing online PPGIS applications.

2. To portray collaborative, real-time Web GIS-based participatory infrastructure that can employ open source geospatial data, standards, software tools, and Web services.

3. To develop and execute a system's prototype (i.e., GeoMeeting) based on the Co-PPGIS model and Web GIS-based framework which encompasses a GIS-based forum, mapping APIs-based spatial component, a notification/feedback component, sub-workflows of information resource providers, and collaborative real-time Web Map sharing Infrastructure that accommodate the stakeholders to share their multiple ideas in real-time scenarios without leaving their place of residence or workspace.

This research primarily encompasses of the working mechanism of real-time collaborative Web map sharing framework that is going to be addressed within a fixed-time period.

1.3. Background and literature review

Increasing importance of the need for an effective public participation in a decision-making process during municipal planning and development is on main focus in this section of study. Through the integration of GIS technologies, involvement of public or local stakeholders in decision-making can become more effective. Public meetings, which are one of the traditional methods of public participation, are integrated in some PPGIS projects to accumulate public ideas, values and preferences [15]. Collaborative use of GIS-based services encompasses the involvement of public and planners in the decision-making process with geo conceptualize a map and accommodate public and planners to build local spatial knowledge and exchange ideas. In order to get instantaneous access and conceptualize the spatial information and participate in decision-making process, collaborative GIS-based services provide opportunities to local stakeholder [16]. An increased public participation can lead to a better and effective decision-making because the processes of decision-making and public participation have a direct relationship, which means that better decision-making processes can also lead to an increase in a user's participation and vice versa.

1.3.1. Rationale on municipal planning and management through existing public participation

In almost every field of life planning process has certain defined goals or objectives just like in developing a small or large scale municipal plan, planning process has some objectives such as, to make planning process accessible, to accommodate in the conveyance or dissemination of ideas, and to support the decision-making process. Participation of public in municipal planning and management, according to traditional methods, includes neighbor notifications, interviews, exhibitions, public meetings/focus group discussions, and public enquires through telephone, letters, mails, fax, or public hearings [17, 18]. In order to disseminate the need of a proposed solution during public meeting, planners and decision makers present their plans through Power Point or point boards which is still considers one of the most commonly used participatory approach [19]. In the western world, public meetings are organized in order to accumulate feedback of public during planning and development-related workflows for effective and better decision-making. For example, in United States and Canada, local governments and many municipalities necessitate a level of participation in their decision-making processes.

Table 1 reveals issues and concerns which are commonly faced during planning and development-related processes in existing practices of public participation. It illustrates or portrays the complete assessment of existing public participation practices related to communication channels, notification, access of information, and exploring spatial data of municipal projects. Li et al. (2006) also disclosed several main issues regarding to traditional public participation practices like inadequate access to the information needed for public input, for exchange of ideas or information and for communication there is a lack of essential or creative platform, restricted awareness mechanisms, and notification channels. Factors like "successfully revealing and educating the public about the program before hearing, proper planning, and

Issues	Concerns	
Notification	Limited means, e.g., newspaper, flyer, etc.	
Communication channels*	Public meetings/public calls/information resource center	Formal/informal presentation
		Open talk with public
		Flat board displays containing preliminary design/model solution
Exploring spatial data	Using hardcopy maps, etc.	
Access of information	Less feedback or public involvement	
	Lack in projects data management	

*Establishing confrontational contact, dominated by higher authority, fix time, feedback lack and accessibility issues.

Table 1. Issues and concerns in existing practice of public participation.

management of meeting, providing an understandable and media-rich demonstration of the issues and organizing a proper follow up" are those factors upon the success of public meetings depends.

According to Meredith et al. [20], for successful public participation, proper and adequate access to information, effective connections to decision-making process and effective tools for getting input into a decision-making process are very essential. Public participation can became better and effective only if a large number of participants easily understand the message and give valuable feedback in a short time-frame.

1.3.2. Rationale on CSCW and groupware

Previous studies related to the depiction and execution of real-time collaborative mapping technologies is still in its stage of growth and development. Although in the last decade, many attempts have been made to the research of developing collaborative PPGIS but despite of this insufficient literature is obtainable in this field [21–24]. The rapid developments in technologies like in GIS, OSGIS, GIT, CSCW, and groupware filed will have a notable impact on the transfer and/or integrate those technologies into collaborative contemporary GIS. Rinner [25] and Li et al. [13] recognize the need to support such technologies that provide a limited way of investigating spatial data or map information collaboratively by inaugurating asynchronous-based geo-referenced mapping architecture.

Baecker [26] defined groupware as information technology used to accommodate people to work together more effectively and efficiently. With the help of the CSCW application or groupware technologies, people in remote places can easily and effectively interact with each other by sharing the documents and files through voice, data, and video links [27, 28]. Using proprietary software approaches, e.g., PCI geo-conference, a few GIS-based tools encompassing groupware and CSCW technologies have been originated. Some attempts have been made to originate simple map sharing applications using open map services. As a result of modern developments in Geographic Information Technology (GIT), that assist large spatial databases, groupware technologies and Web-based GIS, several frameworks that accommodate real-time collaboration were designed and developed [29–32]. Jankowski et al. [33] developed Spatial Group Choice, a spatial decision support framework, to assist the CSCW technique. Churcher and Churcher [34] proposed and developed Group ARC which offers a tool to geographically dispersed people to collaboratively view and explain map/spatial data. Pang and Fernandez [35] designed and developed Real-time Environment Information Network and Analysis System (REINAS) which encompasses functionalities that are helpful in the analysis of geospatial data. In order to support decision-making in Trane China, Trane China SDSS (TCSDSS) was designed by Xiang (2003) by adopting the unified software-development process.

By acquiring "argumentation philosophy," argumap (which is an asynchronous perspective for spatial participation planning, to accommodate group discussions by connecting specific notations to map features) was developed by Rinner [25]. In order to support planning and decision-making processes, SoftGIS was developed which permitted mapping local knowledge and integrating it into urban planning practices [36]. Community Action Geographic Information System (CAGIS), a participatory GIS approach developed by Stewart et al. [37]. Virtual Emergency Operations Center (VEOC) framework was designed for the purpose to provide a collaborative virtual environment that allows connectivity among participants while implementing synchronous, script-driven tests and assumptions [38]. For amplifying collaborative decision-making among geographically distributed people, synchronous collaborative 3D GIS was designed by Chang [6] to assist synchronous collaboration. For participation in community planning, map chat is an online geospatial tool designed at the University of Waterloo. Collaboration in planning and/or emergency management related to decision-making, Rinner [39] recognized OSGIS technologies and OSS-based Web 2.0 concepts (that encompasses n-tiers application client-server architecture, Web mapping tools, Participatory Geographic Information System (PGIS), Web Mapping Services, 3-D GIS technology, CSCW, and Web-based groupware to accommodate consideration in spatial decision-making) that have played an essential role in this regard. The aim of this study is to describe core concepts, design, and technology with an examination of allowing technologies for analyzing and designing a successful real-time map sharing mechanism. This study also narrates a framework development based on a research project that looks into connecting CSCW principles, PGIS, and open source groupware tools with Web-based GIS.

1.4. Summary of closely related research models

Already existing PPGIS' applications or models assessment helped researchers to find limitations of applications' framework and current practices. Three research models which are considered relevant to the present study are discussed below. Rinner [25] introduced the argumentation model, in his model he introduces argumentation maps as an object oriented model for geographically related discussions. As shown in **Figure 1**, it shows the relationships between an argumentation elements/discussion, a geographic reference object/map feature, and user-defined graphic reference objects/sketches [39].

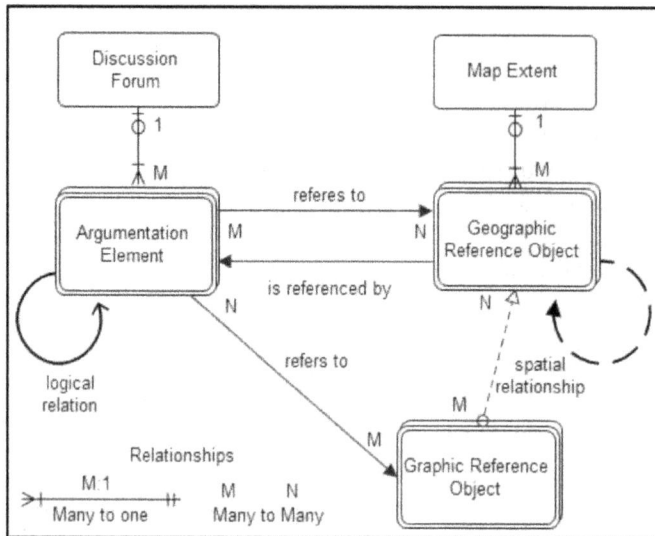

Figure 1. Modified argumentation map model (source: Rinner [39]).

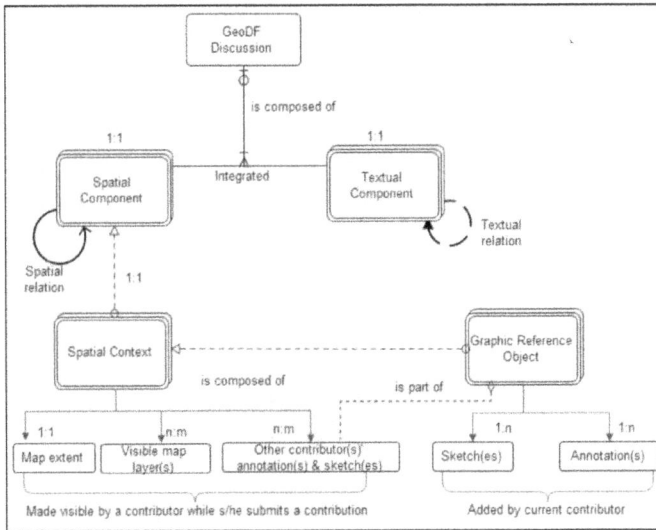

Figure 2. Modified argumentation map model for GeoDF (source: Tang [40]).

The argumentation model object classes have reinforced many-to-many relationships. For example, an object which is geographic can associate many argumentation components and an argumentation component can be associated by many objects that are geographical. Additionally, as shown in **Figure 1**, the objects have their self-relationships to each other of the same class. For example, geographic reference class objects have spatial relations to other object, and argumentation components, class objects can have logical relations to other objects; again, many-to-many relationships are supported [39]. The argumentation model provides an open standard-based prototype with a special focus on the use of standards to confirm interoperability. The discussion component was developed using open source programming languages, i.e., JavaScript and Java applet. The map elements are based on an open source Java API, i.e., Geo Tools and libraries. Same kinds of models were established and acquired by Tang and Hall (2006) and Leahy (2006), but many other technologies were used to design the prototype of research Map Chat and GeoDF. Tang GeoDF model installed an open source-based PHP built in board with commercial-based ESRI Arci MS/spatial server to develop GeoDF prototype. **Figure 2** shows several elements of GeoDF model.

Every conversation is comprised of two main components, i.e., the textual component that is basically related to a respondent's understanding about the shown as text and spatial components that is actually a part of spatial element and is a together term used is geographic features, map extent, location, and spatial relationships embedded in GeoDF discussion threads, which are the thoughts, views, or feedback submitted by a participant via the GeoDF. In other words, the spatial context is primarily comprised of graphic reference objects, i.e., annotation, sketch, and other respondent's annotations and sketches together

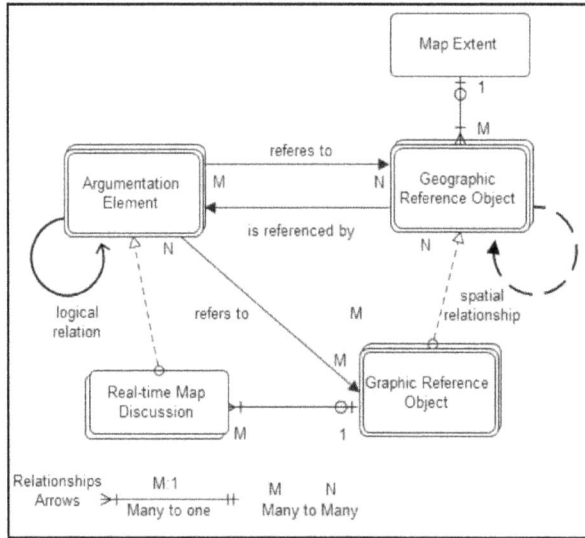

Figure 3. Modified argumentation map model for map chat (source: Hall and Leahy [41]; Rinner [39]).

with other two spatial components, visible map layer, namely the map extent [40]. The Map Chat argumentation model (see **Figure 3**) engages the same classes and objects for spatial and textual relationship in comparison with previously discussed models. A new real-time map discussion class was introduced in this model, which provided the functionality of real-time geo-chatting in connection with every graphic related object. An open source application infrastructure is provided by Map Chat argumentation model. It appoints open standards in relation of the overall system specification and it uses open source coding in PHP and JavaScript based, and it uses a reliable architecture to give the installation of other tools of models [41].

These models have some sort of similarities like to introduce an open standard-based object model, to share a same map extent during discussion, making a spatial relationship with graphic reference objects, and adopting an asynchronous participatory approach for map-based discussion. All three argumentation models allocate structured discussion, about different features of map and geographic related objects, in many geographically meeting respondents to provide an approach of the asynchronous spatial data. For example, the approach with the asynchronous spatial data sharing, it is not possible to find out an argumentation component related to the object of real world simultaneously in different respondents/members. The Map Chat provides geo-chatting discussion functionality with real-time, which cannot be implemented over other two models that used discussion threads with relation objects for geographic referencing. Unifying the chat with discussion elements gives a flexible and a powerful way of managing discussions that are geographically referenced, but participants should train themselves with this function, that is amalgamation, which gets advantage from this reliability.

2. Design modeling of Co-PPGIS

The prosperity of developing and establish a Geospatial enabled Co-PPGIS, for enhancing the ability of people participation in collaborative decision-making during management workflows and municipal planning, most importantly it depends on a brief understanding of firstly the ideas of community participation in management and planning which involves basics ideas of role in participation, amount of community participation, and already existing participation of community at the time of municipal development, planning and management and second important concern is on functional and non-functional requirements, that are identified by existing PPGIS and that is related to research models, which are developed during municipality management to support public participatory processes. It begins with an explanation and overview of a Co-PPGIS idea, which executes the role of a real-time synchronous and asynchronous participatory approach to help the decision makers to make decisions to assimilate people role at the time of a municipal planning process. Some are the information sources and withdraw for the requirements of modeling of an advance Co-PPGIS for planning and management of municipal related projects. Although, it gives an introductory source of information that introduce an idea of advance Co-PPGIS to understand the infrastructure of a Co-PPGIS and to find out the gaps between existing municipal planning processes and possible improvements in Co-PPGIS.

2.1. An idea of advance Co-PPGIS

An idea or concept is a plane, intention image of a specific thing, institution, or a class, and a framework is introduced as a form which gives support to the number of elements and fulfill as a package. Basically, a conceptual framework is a structure of interlinked ideas, which gives support of a certain phenomenon or process to build understanding. Public participation is necessary for the evolution of a country, city, and municipality planning, development, management, and decision-making which will speed up the process of planning. During planning, development, and management of municipality in a city or state, the management of geospatial data remains a challenge. Co-PPGIS gives us a planning and management related spatial and non-spatial information to the decision makers, higher authorities, and government bodies on a basis of real-time basis geospatial Web conferencing infrastructure. In this chapter, the advance Co-PPGIS has focus on municipal projects through developing a GIS-enabled virtual meeting idea. The advance Co-PPGIS framework is showed as five viewpoints, which are shortly discussed below:

Social viewpoint: The first side of social viewpoint in the Co-PPGIS is to highlight and show a name of project which will help stakeholders (see **Figure 4**) to play its role in the related project or matter. Before joining the meeting that will aid the stakeholders to find out the status of all participants submission of user profile, there are some ethics, rules, and values for community in social interaction. Their interaction level rises when the participants join the meeting or session. They exchange their ideas and views, which guide to better decision-making processes for municipal projects.

Geo-spatial viewpoint: This idea links with mixture of time, place, and channels of communication. To address a meeting physically, it is difficult for everyone nowadays. That is why the advanced latest technology provides participants envision the working location. Through GIS

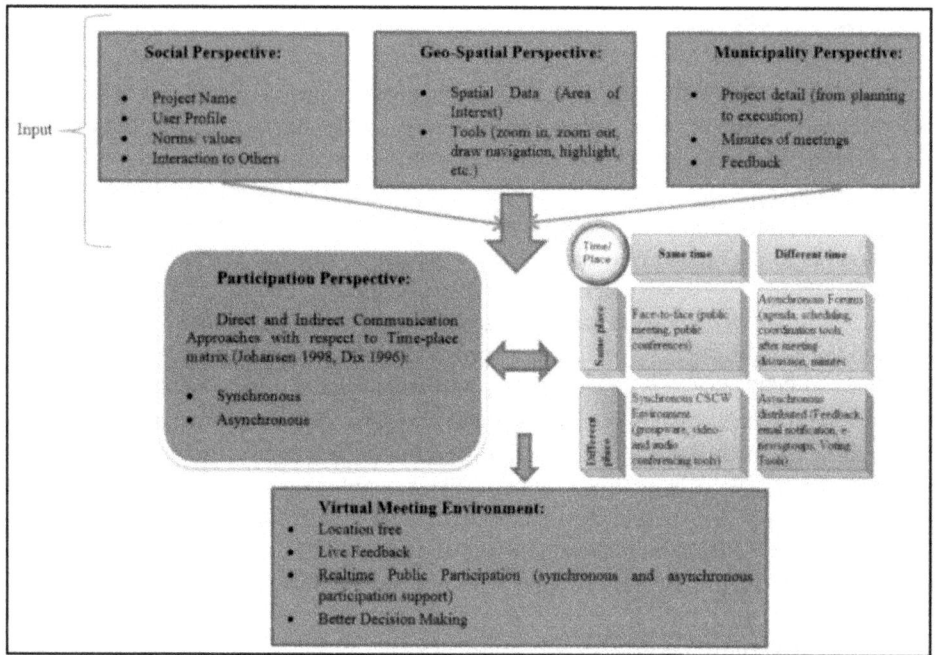

Figure 4. Conceptual framework for proposed collaborative PPGIS.

technology, the advance prototype allows a participant to visualize an area of interest, draw or highlight, navigate on the map, and patch on the map. This is how the participants can seek others for discussion related to analytical issues on any point and the provision of a small point is very essential in any project which is related to municipal.

Municipality viewpoint: In any municipal project, the idea of all information regarding a project at one place is very important. This is how one project from another project differentiates the status in the same domain. The advance collaborative PPGIS has the provision to gather and supervise the data, e.g., planning info, minutes of previous meetings, drawings/maps, feedback form, notification, etc., at one place and a participant can easily get the information at that level which they want. So, a new participant can easily reach the present level after taking information from step one. Public role is very essential in the development of projects, and its importance was not perceived in the last few decades; whereas, community is now playing its essential role for making the decision-making process transparent and better.

Participation viewpoint: This crucial idea is very essential while constructing a collaborative PPGIS. It enhances participant's abilities in the municipality project standard and with their available conditions and time. Through synchronous approach, in which multiple people can see what the other people are doing at the same time without wasting, the second group of participants can share their point of views using drawing tools, mapping. Video chats are the best example of in which everyone can see and understand what other is doing. Stakeholders have indirect communication facility through asynchronous approach in which it is not compulsory to see what the other is saying at the same time. Among stakeholders, filling a feedback form is a good example of indirect communication. The advance PPGIS gives both direct and indirect communication facilities for improving the participation of stakeholders. The best example to fit the advance PPGIS participation viewpoint is the time-place matrix which is categorized according to the spatial and temporal dimensions [42, 43] starts from same time (synchronous) and same place (co-located), different time (asynchronous) and same place, different time and different place (distributed), and same time and different place.

Virtual meeting environment: With the passage of time, technology has become more advance and friendly. The advance Co-PPGIS has a solution in which a participant can easily participate through the electronic meeting facility without appearing physically in the meeting and share his views with relation to project. Participants can do video chat and can drop a message for a specific participant without any restriction. This is how, decision makers can easily involve in any project, which is being developed for a municipality for its effectiveness and efficiency, which will ultimately lead to better decision-making process. In developing countries, resources are minimum and need is maximum, like Pakistan and India. There is a massive need for developing such thing for public, which gives all these facilities, which are mentioned above, to give comfort to decision makers.

Shortly, Co-PPGIS environment is an online meeting procedure for supporting participant's collaborations on geographical information like mapping and imageries, and collecting and sharing data during processes of management. **Figure 5** shows Co-PPGIS virtual meeting workflow processes, service abilities, and to describe situations when its functional capabilities are useful.

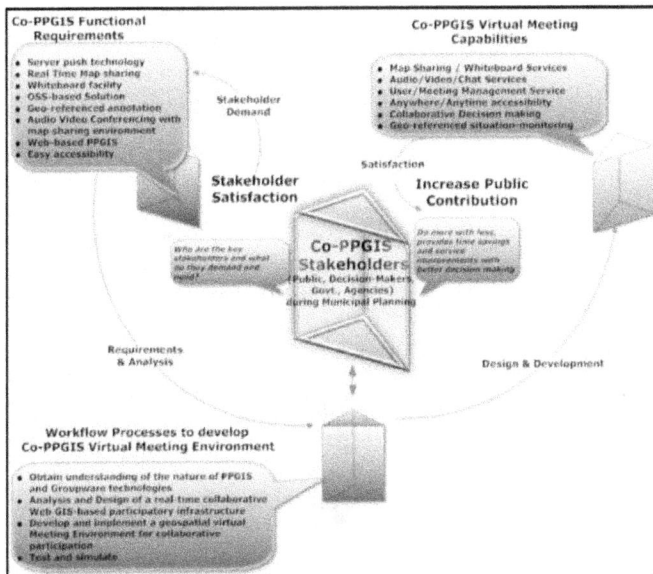

Figure 5. Co-PPGIS workflow processes and service abilities.

This kind of environment allows combination of geospatial data from other sources from Web services and collaborators input through geo-referenced comments. It involves components such as audio/video conferencing, map sharing, geo-referenced textual, real-time chatting and graphical annotation, and user or session management.

2.2. Understanding a Co-PPGIS infrastructure

Co-PPGIS is basically a GIS-enabled collaborative and multi-function, essential meeting participatory infrastructure, which combines different information technology tools to accommodate participation and cooperation activities before public meetings (i.e., the major activities before public meetings is focused on information access, communication, and cooperation of stakeholders), during public meetings (i.e., real-time access to the meetings and their demonstration become paramount tasks), and after public meetings (the focus alters to the demonstration of syntheses of public participation, access to decisions, and receiving of feedbacks). As from above discussion, it is concluded that Co-PPGIS primarily centers on public meetings engaged during the municipal planning and development-related activities (**Figure 6**).

In order to properly and easily understand Co-PPGIS system completely, Co-PPGIS may be categorized as and/or mainly composed of two major application infrastructures. In other words, we can say that recommended Co-PPGIS is basically an amalgamation of two mechanisms of participation, cooperation, and communication between members, i.e., Co-PPGIS asynchronous and Co-PPGIS synchronous. In Co-PPGIS asynchronous participatory environment, Web-based GIS geo-referenced conversation platform and/or GIS Blog techniques are used to accommodate public input and discussion. In Co-PPGIS synchronous participatory map sharing environment, synchronous collaborative GIS processes are executed by applying computer supported cooperative work (CSCW) or groupware application principles. By this, participants can easily share comments, ideas, or suggestions after investigating spatial data by using digital multimedia tools and technologies (**Figure 7**).

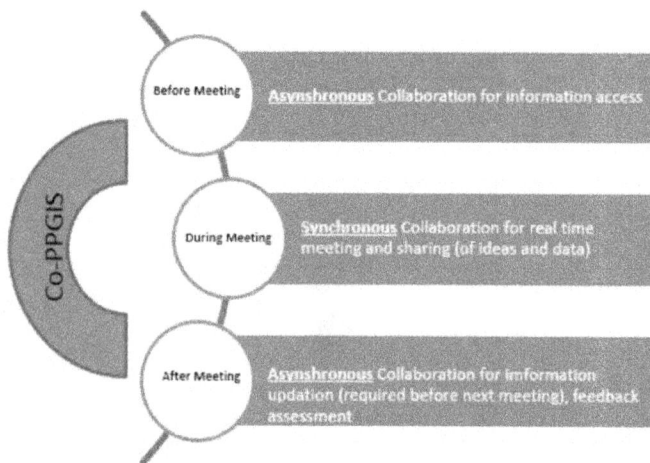

Figure 6. Public meeting scenario at three stages/levels of interaction.

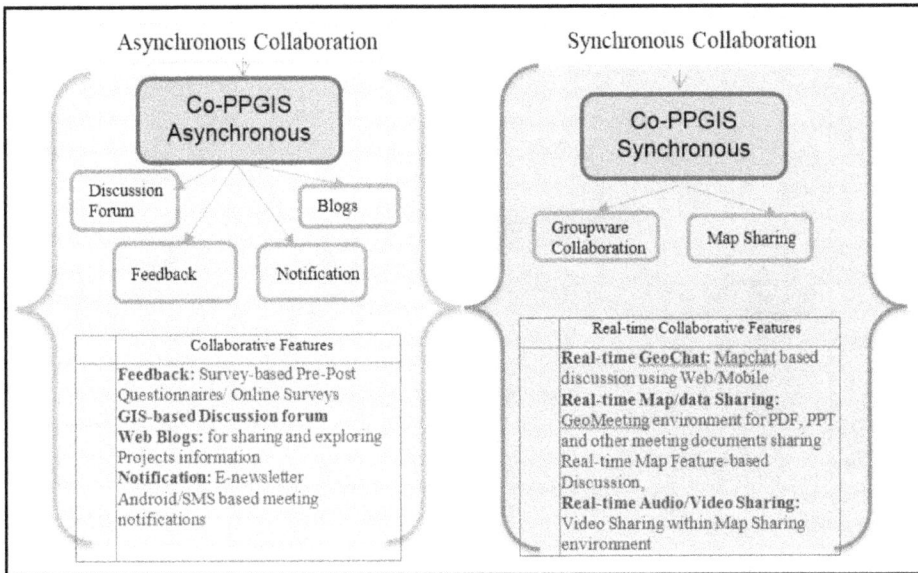

Figure 7. Co-PPGIS virtual meeting components for concept of planning and management.

2.3. Exploring gaps in existing municipal planning practices and possible improvements using Co-PPGIS

Exploring and contrasting of existing PPGIS application's performance is essential or helpful in recognizing the functionality gaps between those collaborative PPGIS applications which organized crucial basis for Co-PPGIS requirement analysis and architectural design. **Figure 8** depicts the research gaps in current or existing communication mediums or participation practices found during the literature review and recommended how the Co-PPGIS contributes to the existing practice in order to increase public participation in municipality planning and development projects. It also explains how the approaches in relation to the proposed/enhanced infrastructure of Co-PPGIS will organize, improve, stimulate, accommodate, and contribute to the existing public participation practices.

Issues and the improvements of these issues through Co-PPGIS are explained in this section. For instance: (1) through or by using Co-PPGIS meeting environment, the issue of inadequate communication, generated due to fixed-time meeting schedules, accessibility issues, lengthy presentations, and open talks with authorities can be accompanied because Co-PPGIS supports anywhere/anytime/anyone accesses with real-time participation support. (2) Through a spatial component of GIS-based platform or through real-time map sharing cooperative component of the Co-PPGIS the issue of inadequate way of investigating spatial data, i.e., using hard copy maps in the meeting sessions because Co-PPGIS increase the degree of public participation along with spatial data investigation during essential meeting sessions. (3) Through meeting scheduling/notifying and/or by the e-newsletter components of Co-PPGIS Blog, the issue of inadequate process of sending notification related to existing municipal

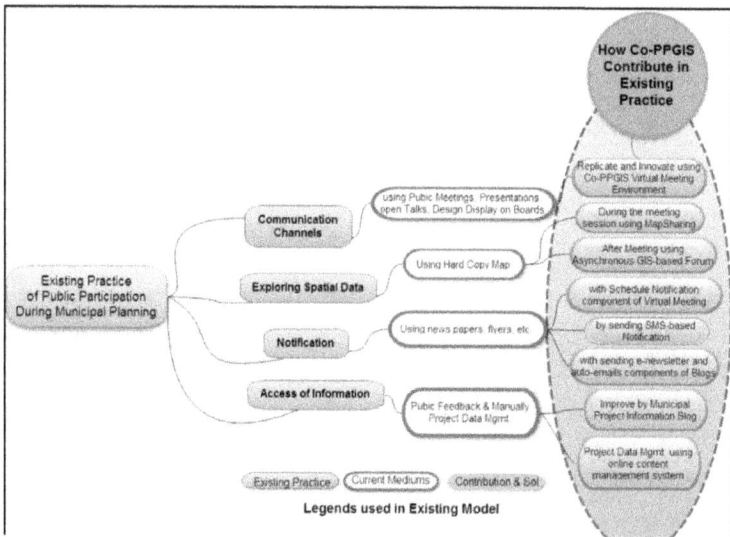

Figure 8. Identifying relation between existing participation practices and the suggested Co-PPGIS.

development projects can be self-regulating/self-operating. (4) Information's access associated to a municipality project's level data can facilitate through project information blog which exhibit the existing or future municipal project's notice detail, minutes of the meeting, presentation, document, location, and all valuable information. (5) Through Co-PPGIS, the absence of support to quick decision-making can be encouraged because Co-PPGIS upgrade or improve public participation or input as well as assist scattered decision makers to work coincidentally on a real-time basis to conclude the decision in timely manners, which eventually diminish the time span of planning and probability of failure.

The upcoming sections demonstrate prototypes' execution of the proposed framework to assist its real-time synchronous participatory procedures that exhibit the innovations to be expected when trying to perceive the concepts established in this research.

3. GeoMeeting service-aligned architecture

The way in to GeoMeeting services can be through login authentication. With direct way in, the user can log in by just choosing a screen name; on the other hand, in login mechanism, user only have to register the user's login for sharing the data and services (see **Figure 9**) provided by Co-PPGIS.

Some of the components-based services provided in the GeoMeeting are explained as follows:

Service for login management: HTML and Web pages are developed to user profile database. HTML used The PHP form very long to sending variables such as name and email address, etc., and receiving results. By using Structured Query Language (SQL), PHP scripts process

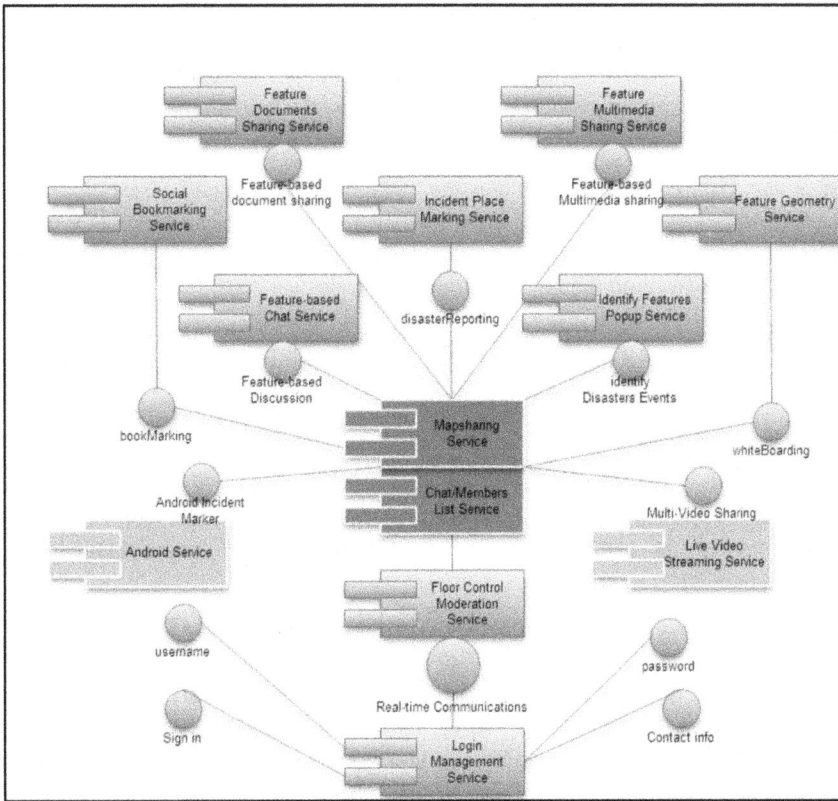

Figure 9. GeoMeeting components-based services.

these input variables and send it to the database for verification. And then, user is moved toward GeoMeeting interface after this process.

Service for floor control: Floor controller is the process which is requested when administrator flicks main toolbars for the purpose to elude the distraction from other attendees during the session of GeoMeeting. The purpose of working behind is that a message is sent through flex service to Blazeds server and shared with other attendees. Once, the IDs of tool bar are collected and stored, i.e., map panning, use of toolbars and scrolling will be flick, and if any participant wants to discuss on the map interface, that user may ask from the moderator by sending a request using JavaScript.

Service for member list: The flex controller saves the user name and sends it into a user list panel which is constructed using flex-based server scripts. When the user signed in or logged in, the widget is shared in html div with other users using Blazeds.

Chat service: In flex and PHP languages, chat object is developed. In flex server-side scripts, when a user types in the chat window, it delivers and shares among other users using Blazeds real-time messaging service.

Service of map sharing: Map sharing idea is designed and got using JavaScript programming languages and Flex. During map sharing, it provides environment, when a user go for the zoom or extent of the map, the value of the extent and zoom is recorded through JavaScript function and send to flex server (Blazeds) which react to all users for map synchronization. In outcome, every person or user can share the same map area on screen.

Android service: User collects and updates the incidence rate using a Java-based application. Information related to textual and multimedia data are changed to xml format and recorded in the database through XML writer and Java Server pages (JSP). XML parser identifies the xml and data are showed on the map through JavaScript.

Feature-based chat service: The feature-based chat element operation and workflow is managed in JQuery and JSP. Popup window process controller gets the information by the user and sends it to the JSP using JQuery. After that, JSP gives the information to the database by an SQL for record and receiving of the message posted. Chat messages are real-time shared in different attendees during meeting using Blazeds Server.

Identify feature popup service: Identify feature popup is constructed to find out the selected elements properties. On elements selection, a JavaScript uses for finding out and showing all feature's elements information inside the popup window through HTML.

Geometry feature service: Geometry feature service is noted in JavaScript languages and JSP. When features are drawn, for example point, line, polygon, circle, annotation, and hexagon on a map and geometry object of the feature are inherited using open layer's JavaScript library. If this library is saved in directory on the Web-server than it can be studied by client browsers, apart from that it can have access from online URL. On the server side, JSP can then find out by reading and parse by a text and then save the geometry into a database of PostGIS. In the return, JSP reads outcomes, for example, attributional information and geometry from database through SQL and xml converted by JavaScript for showing and parsing map shapes.

Another way of developing basic geometry elements in GeoMeeting is through WMS, which is served through GeoServer. The stored characteristics and geometry are declare to GeoServer and can way in or read by browsers through open layer functions and JavaScript. When a new element is developed on a map, JavaScript function take the ID of each element and deliver to the Blazeds, elements are shared to all the other users those are on board.

Feature multimedia sharing service: Feature multimedia sharing service is created to build connection multimedia objects for example images, audio and video files against each geometry shape or incident for the purpose to share rich information among managers. All multimedia files are record on the Web-server; whereas, element information with multimedia objects get through JSP. Multimedia is classified based on type and each category is showed in different GeoEXT panels.

Bookmark service: Bookmark service is constructed to save the extension of the map and important discussions made by different users. When a user clicks on the bookmark, the extent and zoom of the map are saved into the database through JSP Get function.

Live video service: With the live video-based interactive communication service, any person can share audio and/or video while demonstrating his/her ideas on the map. GeoMeeting video service is developed using Flex, Action Script, and JAVA programming languages. The meeting client accesses the RTMP protocol of RED5 streaming video server, for sharing the video among participants. GeoEXT-based video popup sports a very simple user interface so that everyone can focus on the Geo-enabled meetings—not on technology. The participants do not have to install anything (even to broadcast audio), in brief; a single click starts video conferencing among multi-participants.

4. GeoMeeting prototype

In order to aid the Co-PPGIS synchronous participation procedure, which is originally developed and designed to resolve the issues associated with the municipality planning and management, GeoMeeting Prototype is executed as a proof of concept. GeoMeeting Prototype was developed and designed for effective geo-cooperation among National Society, government, local, and international NGOs. GeoMeeting prototype is basically a Web-based geospatially enabled conferencing system that accommodates synchronous and real-time amalgamation of data from different sources through Web map services like APIs, and supports the amalgamation of local knowledge demonstration by meeting participants. It also supports real-time map sharing, geo-referenced map notations, geo-chatting, user and meeting management for accommodating conversations among multiple users that are geographically located at different places. GeoMeeting is developed from scratch, amalgamating the technologies of open layer and flex technologies, having associated step by step development processes (that mean limitations discovered during the first version of prototype is enhanced in the next version of the development).

GeoMeeting system which is Geo-enabled comprises the following capabilities:

- As all the multiple users and participants in a GeoMeeting can sight the same geo-referenced map simultaneously that is why it is called geo-enabled GeoMeeting system.

- In order to undertake synchronous conferencing, the GeoMeeting server application employ a push technology procedure like real-time instantaneous messaging are typical examples of push services.

- GeoMeeting provides Real-Time Map sharing among multi-users or participants.

- GeoMeeting is provided by geo-referenced pointer with a purpose of pointing at the shared view of map.

- With the aid of whiteboard facility, multiple users or participants can produce geometry-based incidents

- GeoMeeting provides the opportunity of proper handling of maps (like modifying layers, map scale, and its position) to participants and users. It is very easy to rotate or change map view among different base map layers like street map, satellite, hybrid, and terrain in GeoMeeting prototype.

- In GeoMeeting, participants or multiple users can easily produce and share geo-referenced notations.

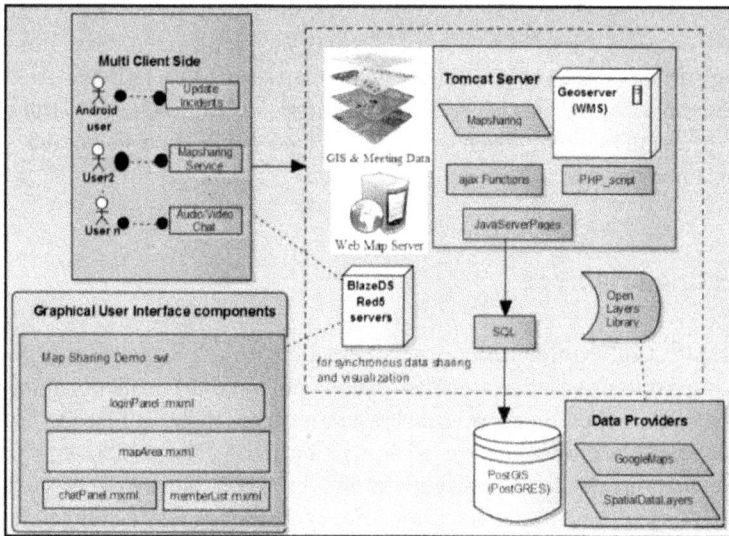

Figure 10. Conceptual architecture of the GeoMeeting system.

- Construction and installation of GeoMeeting prototype, a Web-based client-server architecture, is very easy; we just need to plug and play.

- Through the use of any browser like Chrome, Opera, Internet Explorer, or Firefox, GeoMeeting prototype provides the opportunity of easy accessibility of the main interface of a prototype to the users.

- Online map sharing application is depicted using open source technologies, APIs and programming languages like Flex SDK, MXML, Adobe blaze DS, Java Script, Action Script, and open layer API, etc.

- GeoMeeting application is considered extremely useful during collaborating decision-aimed events such as emergency response, disaster management, and urban planning activities because GeoMeeting is a live conference technology.

GeoMeeting has myriad of capabilities but its operational status is still in its progressive stage. **Figure 10** demonstrates a conceptual architecture of the GeoMeeting system.

The upcoming section's discussions are based on the execution of different versions associated with the GeoMeeting prototype development.

5. Walkthrough of GeoMeeting prototypes

This section explains functionality requirements and enabling technologies of three GeoMeeting prototype's design.

5.1. GeoMeetingV1

5.1.1. Key features

Participants may visit the log-on page for GeoMeeting using a standard Web browser such as Mozilla Firefox 13+. Once the log-on page is displayed, the person can enter a user name and connect the GeoMeeting environment. After the meeting session is entrenched, the GeoMeeting elements will charge its default interface as shown in **Figure 11**.

Depending on your connection speed, the loading of Web services from different sources may take only a few minutes or sometimes take few seconds. The GeoMeeting component provides the following key functions: *Caption A* illustrates two pointers. Black pointer will activate the geo-referenced pointer on the GeoMeeting component that has a means of "gesturing" at the map to other participants and helps in highlighting an study area issue which can be seen by all participants on a real-time basis, whereas the white pointer will deactivate the pointer on the GeoMeeting component to other participants who can be useful for single user/moderator work. All participants map will relocate as of the moderator, and participants can have discussions over there. *Caption B* illustrates a search toolbar. Search option provides the way to find a specific place of interest. The textual/graphical comments may be included at a see-through map layer and act as a shared whiteboard. The features can be in the configuration of points, lines, and polygons. These configurations are geo-referenced which means that they will be scaled as the map display is zoomed and extent which show outcomes in no misrepresentation of the configuration. The geo-referencing toolbar is constructed for the purpose to show the real-time supportive geo-referenced based configuration on map. The annotation tools are

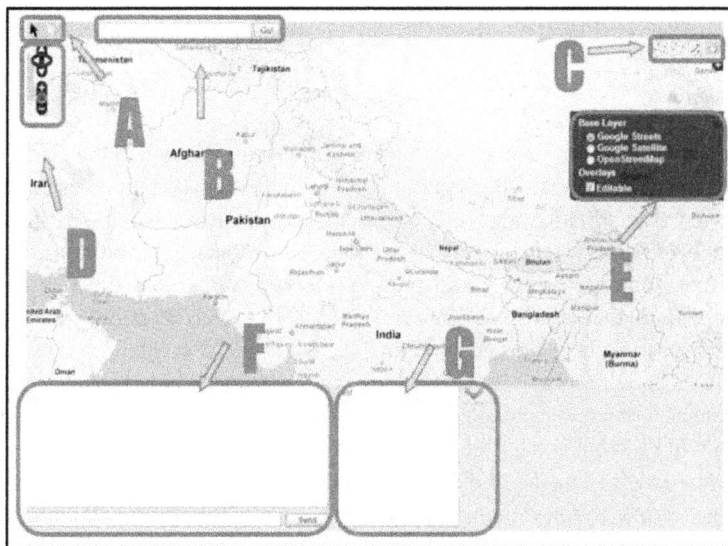

Figure 11. GeoMeeting interface.

organized in a toolbar. To use the annotation tools, click on the appropriate icon on the toolbar and then click on a location on the map or map feature. Add a point, draw a line or draw a polygon. *Caption C* illustrates the textual/graphical notations. A standard toolbar is provided with map zoom-in, zoom-out, and pan functions as shown in *Caption D*. This is basically a typical control in real-time map sharing environment, which allows the user to scroll curser for zooming, left or double-clicks the left mouse button for re-centering the map image to the area where you double-click on the map. *Caption E* is a base layer switcher which is represented by giving and allowing participants to switch between different base maps and including maps given by any Web Map Services (WMS), Open Layer, and the Open Street map layers recording textual and graphical configuration. *Caption F* represents the real-time chatting; it allows any participant in the meeting session to exchange textual information. Chat window panel is developed for sending and receiving messages to other people, as though they are all in the same room, looking at the same map view. In other words, it provides a way for adding remarks linked with the spatial context. The list of participants attending the meeting session is provided through an embedded window panel, shown in a *caption G*. The big down-side blue arrow is used to hide/display the real-time chatting and participants list interface.

5.1.2. Technology

Prototype Web client interface is executed in JavaScript, Adobe Flex, and Action Script. The clients communicate with each other with the application server which is known as Tomcat and real-time messaging server which is Blazeds using a typical set of protocols, i.e., HTTP. The construction of this GeoMeetingV1 makes use of the following technologies such as Web servers (Tomcat), Web map server and tools (GeoServer), Server-side programming (MXML), Client side programming (HTML, JavaScript, Dynamic HTML, ActionScript), Real-time Messaging Server (BlazeDS), and APIs: Google map API V2 and Open layer.

5.1.3. Discussion

The GeoMeeting elements are basically the technological breakthrough in the field of geo-information management. The GeoMeeting is an evolution of map sharing component built previously based on collaborative PPGIS framework, which supports innovative map sharing component technology for better decision-making. There are some issues that need to be addressed and minimized in the upcoming versions of the advanced prototype. At first step, the mouse pointer was used with pixel value that uses screen coordinates for movement of pointer on a map element. But, when it comes to higher or lower resolution, the technique of pointer displays a wrong geographic area. This issue was resolved by justifying the map panel to be left aligned as the problem remained same in center and right alignment, the screen coordinates to initialize from a top-left corner which will be same in all screen sizes, but another problem arises that when mapsharing component was left aligned and displayed on a bigger screen, there is a lot of vacant/empty white space generated on the right-hand side of the screen. This issue will be resolved in the GeoMeeting V2 in which the mouse pointer is synchronized with map coordinates transformation technique instead of using screen coordinates, and participants get the same geographic area as of the moderator.

To make the interface eye catching, the map should be center aligned, but when we center align the map, flash div was moving toward a problem. Flash div is just like a receptacle in the programming language which saves different type of codes in it. Map chat panel and user list are all in different div's and all in the specific percentage, i.e., 100%; when we center align, the flash div's does not visualize the map, chat, and user panel properly. Its solution is sorted out in GeoMeeting V2 by assigning pixel value to chat, and users list div same as of map div in order to visualize it properly. Editing toolbar is used for the demarcation of point, line, and polygon, and a hand tool is used for map panning. Initially, it creates a problem as we select any of them (point, line, polygon, and hand) it did not select properly. In fact, the map container (window) was placed over the editing toolbar, that is why it was creating a problem. As a solution, the editing toolbar's z-axis position was changed (by increasing its z-index value) on the map so that selection of the editing tools can be developed properly and easily. Another issue in the GeoMeeting V1 is that when tools are selected from the editing toolbar and we draw any feature on a map it was not drawn correctly. Many times elements got stuck with the map panel and not allowed it to draw. At the time of development, this issue was seen because of the conflict in different versions launched at different timings. As a solution, old version 2.10 open layer library was replaced with version 2.12 until it was identified and detached from the open layer libraries.

6. Using mock-up case scenarios illustration of the prototype: GeoMeeting implementation in planning and management related activities

GeoMeeting prototype has been used in myriad different fields of studies like public security, crime mapping, disaster occurrences, layout frameworks for preparedness and emergency responses during a disaster, environmental and resources and local government. The section below is comprised of the demonstration of some points using the scenario-based discussion (i.e., the instances come from incidents or cases developed in different regions of the world).

6.1. Scenario 1: GeoMeeting: how well municipalities are meeting the need for parks

As parks are crucial to communities, because parks provide opportunities to people or public for exercise and experience nature which are paramount for physical and mental well-being of humans, it helps in revitalizing local economies; so there is a dire need to estimate park areas and raise municipality residents' living standards. For recognition and generation of easy access to basic essentials of life, such as national parks, green infrastructure, recreation, etc., Government bodies are primarily accountable or answerable because they maintained the available park data but they are not sufficient to fulfill the challenges of data handling and sharing. Therefore, a Web-based GeoMeeting system has been launched to handle and share real-time data, support cooperative meeting sessions at regular intervals, and provide a set of tools like point that helps to recognize park areas and unrevealed base map information about the infrastructure. Video- or audio-based map sharing and Geo-teleconference have been made easy for decision makers to make a decision quickly. Every sort of editing, made by presenter, associated to park marking, will be displayed to all users simultaneously for collaborative decision-making. The presenter can spot a place, add comments, and interpret information related to: park access analysis at city level, complete information and data about every city park, identifying the areas where need of parks is most essential, and recognizing which improvements would provide the greatest advantage to local park system. This sort of

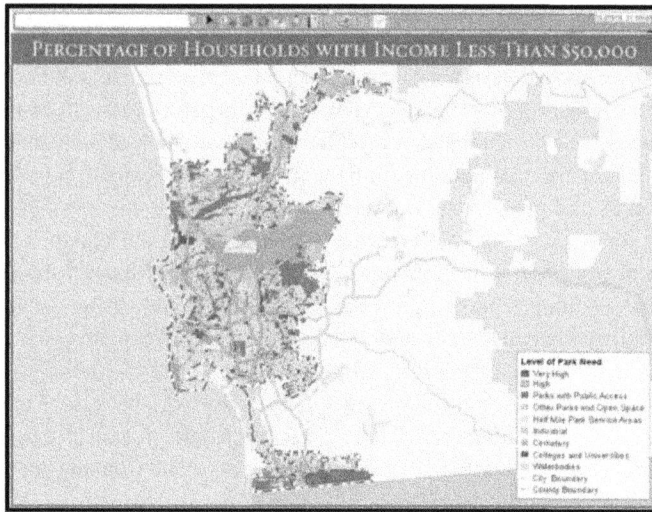

Figure 12. GeoMeeting for municipality parks mapping.

information is accumulated in a database and can be easily recovered. **Figure 12** represents the identified park areas.

Through GIS land use survey techniques, parks are easily determined. In this scenario, android-based techniques are used for collection of park points'. As parks can be easily inspected in high resolution imagery, therefore, high-resolution satellite imagery is used for discussion as a base map. GeoMeeting environment offers the platform for stakeholders to share and execute their views related to area's enhancement by offering better utilities, facilities, and living standards to residents.

7. Usability testing/evaluation

Evaluating the usability testing of GeoMeeting prototype using a case study scenario is helpful to make it effective and usable. A brief summary of steps performed during the usability evaluation of the prototype is discussed as follows.

The evaluation will organized in three parts: (1) a pre-questionnaire comprised of queries related to the user's background, their experience of other Web GIS applications, their computers expertise, and GIS knowledge. (2) Second part of evaluation is the actual user's interaction with the GeoMeeting system using the analytical method with the help of TeamViewer, Session Cam, and Google Analytics tool, which is easy to use, free, and user-friendly usability-evaluating tool that provides a comprehensive set of Website data tracking and analysis tools. By using the TeamViewer and Session Cam recording components, it is possible to collect highly detailed and useful information about the actual usage of the GeoMeeting Website and its components.

Data elements are valuable for evaluating the usability as well as estimating the degree of public input during the process of participatory planning and effective decision-making. A Web-based feedback component was developed to evaluate and measure the usability aspects collected using pre-post questionnaires and Web analytic tools, i.e., TeamViewer, Google Analytics, and Session Cam. (3) Finally, the users were asked to fill out a post-questionnaire comprised of queries related to usefulness, ease of use or interactivity of using GeoMeeting interfaces in order to collect feedback concerning the usability of the system.

8. Concluding remarks

Co-PPGIS, a Web-based geospatially enabled conferencing system, assists a real-time participation to facilitate and improve public participation for collaborative decision-making which will bring fundamentally more understandability in any system. This Web system provides real-time amalgamation of data from different sources through Web map services, such as APIs, and supports the amalgamation of local knowledge expressed by meeting participants. In order to aid the Co-PPGIS synchronous participation procedure, which is originally developed and designed to resolve the issues associated with the municipality planning and management, GeoMeeting Prototype is implemented as a proof of concept. GeoMeeting Prototype framework facilitates any sort of e-governance, management, and emergency scenarios (e.g., municipal planning, forest management, urban sprawl, lands state, crime mapping, disaster response, etc.) related to collaborative decision-making and provides an effective, valid, and see-through system in which all the discussion and recommendations between authorities/participants are conserved in the database and can be viewed anytime to know the irresponsibility of even a common person to some authority handling the entire situation. The GeoMeeting is an evolution of map sharing component built previously based on Collaborative PPGIS framework which accommodate effective and better decision-making through its innovative map sharing component technology.

The infrastructure of GeoMeeting was established on several components-based services such as Login Management, Floor Control, Map sharing, Android, feature-based chat, feature popup service, Geometry and Multimedia Sharing feature Services, Bookmark, and Live Video Services. Registered users can have direct access to GeoMeeting through login authentication. The component also includes chat facility, drawing specific location (point, line, and polygon/area); base layer switcher for better understanding of map and search any area of interest; synchronously. These components-based services make it an effective and efficient platform for information/data sharing. Previously, teleconferencing was the only medium used during emergency management planning, but the drawback for teleconferencing was the absence of any geo-collaborative console, i.e., map sharing. GeoMeeting provides real-time geo-collaboration; which improves accuracy and efficiency as well as saves cost and time of the emergency management organization. Consequently, this Co-PPGIS framework-based GeoMeeting provides an interactive interface to have Geo-enabled collaborative participatory discussion platform among decision-making authorities and common people.

Author details

Muhammad A. Butt[1,2*], Syed Amer Mahmood[1], Javed Sami[1], Jahanzeb Qureshi[1], Muhammad Kashif Nazir[1], Amer Masood[1], Khadija Waheed[1] and Aysha Khalid[1]

*Address all correspondence to: m2butt@ryerson.ca

1 Department of Space Science, University of the Punjab, Pakistan

2 Department of Civil Engineering, Ryerson University, Toronto, Ontario, Canada

References

[1] Densham PJ, Armstrong MP, Kemp KK. Collaborative spatial decision making - scientific report for the initiative 17 specialist meeting. NCGIA Technical Report 95-14. Santa Barbara, California: NCGIA; 1995

[2] Armstrong MP. Perspectives on the development of group decision support systems for locational problem-solving. Geographical Systems. 1993;1(1):69-81

[3] Chung G, Jeffay K, Abdel-Wahab H. Dynamic participation in computer-based conferencing system. Journal of Computer Communications. 1994;17(1):7-16

[4] Al-Kodmany K. Visualization tools and methods in community planning: From freehand sketches to virtual reality. Journal of Planning Literature. 2002;17(2):189-211

[5] Brail RK, Klosterman RE. Planning Support Systems: Integrating Geographic Information Systems, Models, and Visualization Tools. Redlands, California: ESRI Press; 2001

[6] Chang Z. Synchronous collaborative 3D GIS with agent support [PhD thesis]. Ryerson University, Toronto, Canada. 2010

[7] Huang B, Jiang B, Lin H. An integration of GIS, virtual reality and the Internet for visualization, analysis and exploration of spatial data. International Journal of Geographic Information Science. 2001;15(5):439-434

[8] Klosterman RE. Planning Support Systems. Redlands, California: ESRI Press; 2001. pp. 1-23

[9] Roseman M, Greenberg S. Group kit: A groupware toolkit for building real-time conferencing applications. In: Proceedings CSCW92, ACM. 1992. pp. 43-50

[10] Evans A, Kingston R, Carver S, Turton I. Web-based GIS used to enhance public democratic involvement. In: Geocomp '99 Conference Proceedings, Mary Washington College, Virginia, USA, Jul 27-28, 1999. 1999

[11] Jankowski P, Nyerges T. GIS supported collaborative decision making: Results of an experiment. Annals of the Association of American Geographers. 2001;91(1):48-70

[12] Jankowski P, Nyerges T. Toward a framework for research on geographic information-supported participatory decision-making. URISA Journal. 2003;15(1):9-17

[13] Li S, Guo X, Ma X, Chang Z. Towards GIS-enabled virtual public meeting space for public participation. Photogrammetric Engineering and Remote Sensing. 2007;**73**(6):641

[14] Ventura S, Niemann B Jr, Sutphin T, Chenoweth R. GIS-enhanced land-use planning. In: Community Participation and Geographic Information Systems. London: Taylor and Francis; 2002. pp. 113-124

[15] Hopkins LD, Twidale M, Pallathucheril VG. Interface devices and public participation. In: Proceedings of the 3rd Annual PPGIS Conference of Urban and Regional Information Systems Association, Madison, United States. 2004. pp. 71-83

[16] Healey P. Collaborative Planning: Shaping Places in Fragmented Societies. London: Macmillan; 1997

[17] Kingston R. Web-based PPGIS in the United Kingdom. In: Craig WJ, Trevor TM, Weiner D, editors. Community Participation and Geographic Information Systems. London: Taylor & Francis; 2002. pp. 101-112

[18] Li S, Chang Z, Yi R. GIS-based internet notice board to facilitate public participation in municipal developments. In: Proceedings of the 20th ISPRS Annual Congress, July 12-23, 2004, Istanbul, Turkey. 2004. pp. 269-274

[19] Lowndes V, Pratchett L, Stoker G. Trends in public participation: Part 1—citizen's perspectives. Public Administration. 2001;**79**(2):445-455

[20] Meredith TC. Community participation in environmental information management: Exploring tools for developing an impact assessment preparedness program, a report from Canadian environmental assessment agency. 2000. Available from: http://www.ceaa.gc.ca/015/0002/0016/print-version_e.htm [Accessed: April 25, 2004]

[21] Grabot B, Letouzey A. Short-term manpower management in manufacturing systems: New requirements and DSS prototyping. Computers in Industry. 2000;**43**:11-29

[22] Hunkeler D. A decision support system for life cycle management. In: Proceedings of Eco-design '99: First International Symposium On Environmentally Conscious Design and Inverse Manufacturing. Tokyo, Japan: IEEE Computer Society, Technical Committee on Electronics and the Environment. Feb 1-3, 1999. 1999. pp. 728-732

[23] Hsieh MD. A decision support system of real time dispatching in semiconductor wafer fabrication with shortest process time in wet bench. In: Semiconductor Manufacturing Technology Workshop. 2002. pp. 286-288

[24] Marinho J. Decision support system for dynamic production scheduling. In: Proceedings of the 1999 IEEE International Symposium on Assembly and Task Planning. Porto, Portugal: ISATP; 21-24 Jul 1999. pp. 424-429

[25] Rinner C. Argumentation maps—GIS-based discussion support for online planning [PhD dissertation]. University of Bonn, Germany. 1999

[26] Baecker RM. Reading in Groupware and Computer Supported Cooperative Work: Assisting Human-Human Collaboration. San Francisco: Morgan Kaufman; 1993

[27] Abdalla R, Li J. Towards effective application of geospatial technologies for disaster management. International Journal of Applied Earth Observation and Geoinformation. 2010;**12**:405-407

[28] Antunes P, Zurita G, Baloian N. A model for designing geo-collaborative artifacts and applications. Groupware: Design, Implementation, and Use. 2009;**5784**:278-294

[29] Boulos M, Warren J, Jianya G, Peng Y. Web GIS in practice VIII: HTML5 and the canvas element for interactive online mapping. International Journal of Health Geographics. 2010;**9**:14-26

[30] Churcher N, Churcher C. Real-time conferencing in GIS. Transactions in GIS. 1999;**3**(1): 23-30

[31] Dragicevic S, Balram S. A web GIS collaborative framework to structure and manage distributed planning processes. Journal of Geographical Systems. 2004;**6**:133-153

[32] Jones RM, Copas CV, Edmonds EA. GIS support for distributed group-work in regional planning. International Journal of Geographical Information Science. 1997;**11**(1):53-71

[33] Jankowski P, Nyerges T, Smith A, Moore TJ, Horvath E. Spatial group choice: A SDSS tool for collaborative spatial decision making. International Journal of Geographical Information Science. 1997;**11**(6):577-602

[34] Churcher N, Churcher C. Group ARC—A collaborative approach to GIS. In: Proceedings of 8th Annual Colloquium of the Spatial Information Research Center, University of Otago, New Zealand: Spatial Information Research Centre (SIRC) 1996 July 9-11. 1996. pp. 156-163

[35] Pang A, Fernandez D. REINAS instrumentation and visualization. In: Proceedings, OCEANS '95. MTS/IEEE. Challenges of our changing global environment, San Diego, Oct 9-15, Cartography and GIS, GeoVista Centre; 1995. pp. 1892-1899

[36] Rantanen H, Kahila M. The SoftGIS approach to local knowledge. Journal of Environmental Management. 2009;**90**(6):1981-1990

[37] Stewart EJ, Jacobson D, Draper D. Public participation geographic information systems (PPGIS): Challenges of implementation in Churchill, Manitoba. Canadian Geographer/ Le Géographe Canadian. 2008;**52**(3):351-366

[38] Fiedrich F, Burghardt P. Agent-based systems for disaster management. Communications of the ACM. 2007;**50**:41-42

[39] Rinner C. Mapping in Collaborative Spatial Decision Making. Collaborative Geographic Information Systems. Hershey, PA: Idea Group Publishing; 2006. pp. 85-102

[40] Tang T. Design and implementation of a GIS-enabled online discussion forum for participatory planning [M.Sc.E. thesis]. Department of Geodesy and Geomatics Engineering Technical Report No. 244, University of New Brunswick, Fredericton, New Brunswick, Canada. 2006. 151 p

[41] Hall GB, Leahy MG. Internet-based spatial decision support using Open Source tools. In: Balram S, Dragicevic S, editors. Collaborative Geographic Information Systems. Hershey: Idea Group Publishing; 2006. pp. 237-262

[42] Johansen R. Groupware: Computer Support for Business Teams. The Free Press; 1988

[43] Dix A, Finlay J, Abowd G, Beale R. Human-Computer Interaction. 2nd ed. Prentice Hall; 1998

Chapter 3

Identification of Karst Forms Using LiDAR Technology: Cozumel Island, Mexico

Oscar Frausto-Martínez,
Norma Angelica Zapi-Salazar and
Orlando Colin-Olivares

Additional information is available at the end of the chapter

http://dx.doi.org/10.5772/intechopen.79196

Abstract

Morphological relief analysis allows the identification of geomorphological forms and cartographic-environmental studies make extensive use of the medium (1:50,000) and large scale (1:250,000), where the topographical contrast is evident. However, at a detailed scale (<1:20,000) and for territories where the contrast of relief does not exceed 10 m in height, the morphological analyses must be adapted accordingly, because they contribute information to altimetry studies and to the topographic configuration of units. Thus, through visual interpretation and manipulation of high-resolution topographical LiDAR data from Cozumel Island, a relief analysis is presented at a detailed scale for the purpose of recognizing the geomorphological units of karst origin, using altimetry and slope cartography, digital models of elevation, and shading that permits the identification of 109 new exokarstic doline and uvala formations.

Keywords: GIS, karst, relief, island, Caribbean

1. Introduction

Geologically, the Island of Cozumel's basement is calcareous from the upper Tertiary period and reef from the Holocene period, which, accompanied by tropical climatic elements, give rise to dissolution karst formations. The study uses records of uvalas and cenotes in the area of study, which were obtained through a record from touristic service providers and from the community, as well as from scientific articles [1, 2]; however, there is no updated and complete database. For the purpose of contributing to the identification of karst forms (dolines, uvalas, and poljes-locally called "cenotes" and "rejolladas"), altitude mapping and digital models of elevation, shading, and slope generated with LiDAR (light detection and ranging) data were used for identification and cartography.

The word LiDAR is an acronym for the term *LIght Detection And Ranging*, that is to say, detection and measurement of light. This technique is currently becoming a basic tool in studies based on topographical analysis and precision of the information base [3–5]. The use of LiDAR products has greatly improved and has a significant influence in the earth science disciplines [6, 7]. Its usefulness is emergent in shallow reliefs and those with little altitudinal difference.

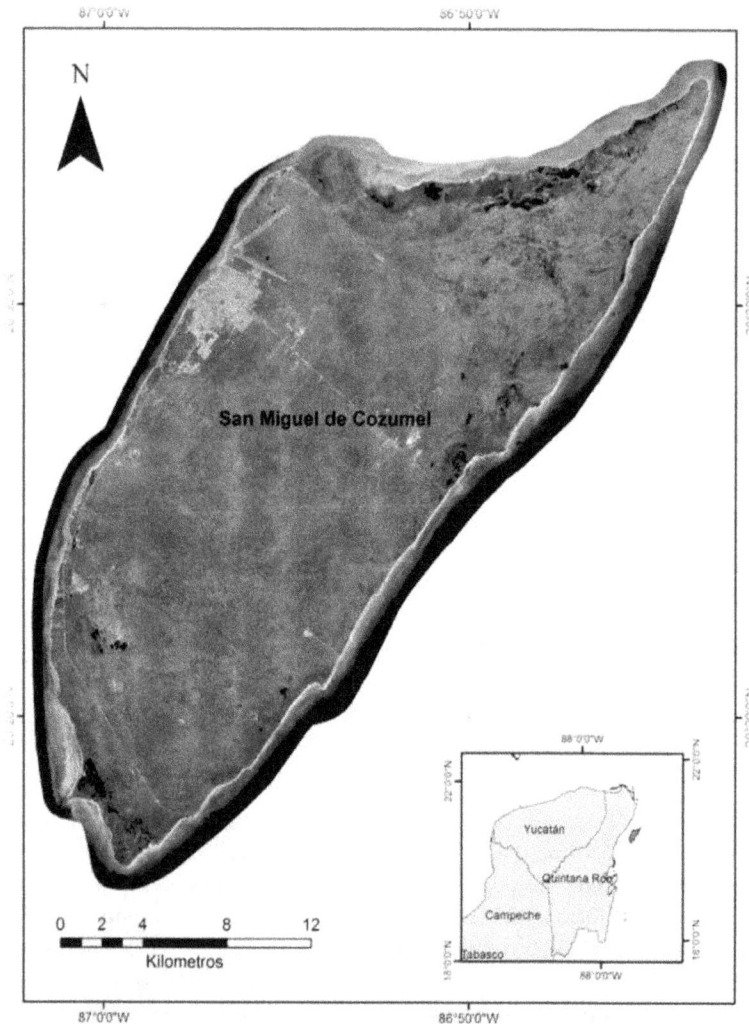

Figure 1. Location of Cozumel Island in Mexico.

The Yucatan Peninsula has an area of 39,340 km² and is located in southeast Mexico's. The most outstanding structural features of Yucatán are the sinkhole region and the aligned islands of Cozumel and Mujeres. The altitude of Yucatán not exceeding 300 m elevation dominates. Previous studies have recognized that different types of karstic depressions abound in the vast plateaus of the northern and eastern Yucatan Peninsula, and there are also extensive systems of caves and caverns in the entire landscape [8]. Climatic subtypes are warm and humid with summer rains and warm and humid with summer and winter rains [9]. Cozumel is part of the Yucatan peninsula with sedimentary rocks formed on a wide platform. The core drillings indicate that the island is formed from reef sediment with a thickness of 100 m or more, which dates from the Oligocene and Quaternary Epochs [10].

Cozumel Island pertains to the state of Quintana Roo, located at 20° 28′ N, 86° 55′ W (**Figure 1**). Cozumel Island faces the coast of the Yucatan Peninsula in the Caribbean Sea, approximately 16.5 km to the east of the Yucatan Peninsula, in the zone of the Northwest Caribbean [11]. The climate is hot and humid with abundant rainfall in the summer. The average annual temperature is 25.5°C and precipitation reaches 1504 mm per year. Cyclones have an important effect, increasing the amount of rainfall in the summer [12].

The relief of the island does not exceed 15 m of altitude above the sea level and the cartographic representations have been developed with data at 1:75,000 and 1:50,000 scale [1, 13]. The presence of dolines and uvalas with exokarstic forms has been reported, but the majority have been characterized over time, due to the fact that they do not exceed 50 m in diameter [14–16], which makes it necessary to recognize other techniques that aid the identification of exokarstic forms. For that reason, this study employs LiDAR modeling for the purpose of recognizing doline and uvala forms, enriching the list of these relief forms, and indicating the areas with higher density of exokarstic forms.

2. Methodology

To achieve the objectives, the following were necessary (**Figure 2**):

Phase 1. Revision of the inventory of exokarstic forms (caverns, dolines, uvalas, and poljes – "cenotes" and "rejolladas") reported in Cozumel Island [1, 2, 13–17], as well as the reports from the association of speleologists MAYAB AC [18] were consulted, 37 of the reported karst units were recognized in these studies, and a georeferenced database was also generated for the purpose of making the information available for reference.

Phase 2. Construction of a unified LiDAR mosaic of the terrain of Cozumel Island. To do so, the data from point clouds in 32 TIFF information files generated by the ALS 40 system and an information area of 62,556.27 km² [19] were revised and corrected; the data resolution is 5 m for the X and Y axes and 15–20 cm for the Z axis. The resolution scale is 1:10,000 for each cartographic model.

Phase 3. The application of filters for the elaboration of morphometric digital models of elevation, slope, and shadows. The digital model of elevation was derived from the simple method

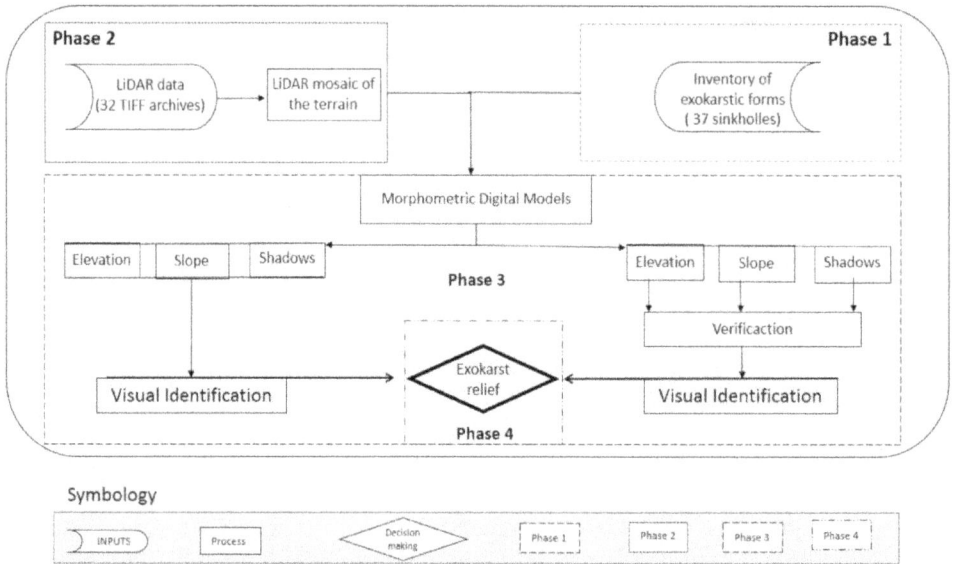

Figure 2. Flow diagram of the methodology.

of nearest neighbor interpolation, rated at 0.05 cm elevation, recognizing up to 16 m of altitude in the area of study [20]. For the terrain shading method, an azimuth of 272° and an elevation of 30° was used, and for the slope map, eight categories of classification were set manually, where the highest category is >40° [21].

Phase 4. Analysis of the distribution of units in the inventory and their contrast with the new models of identified forms. In each one of the models, 109 dolines and uvalas were visually identified, using the high resolution models and the contrast of the altimetric data, slope, and shading. Likewise, the information was contrasted with the 37 units reported in the previous studies. Furthermore, field trips were carried out during the months of June and July 2014 to confirm the occurrence of the dolines and uvalas, where the cartographic prospection at 1:10,000 scale was verified.

3. Results and discussions

In the identification of the exokarstic forms, the primary input used was a mosaic of LiDAR images of terrain with a resolution of 5 horizontal m and 0.50 vertical cm, to which later a shade filter with azimuth 272° and height 30° is added (**Figure 3**). Here, dolines and uvalas with diameters between 20 and 125 m and up to 9 m of depth can be identified, as well as basic linear details of the relief. In the mosaic, one can see the geological structure of the quaternary period (old coastal mountain ranges, dunes, and marine terraces).

By applying an altimetric filter with a vertical difference of 0.50 cm, karstic formations are distinguishable. For said identification, the distribution of pixels based on their altitude values

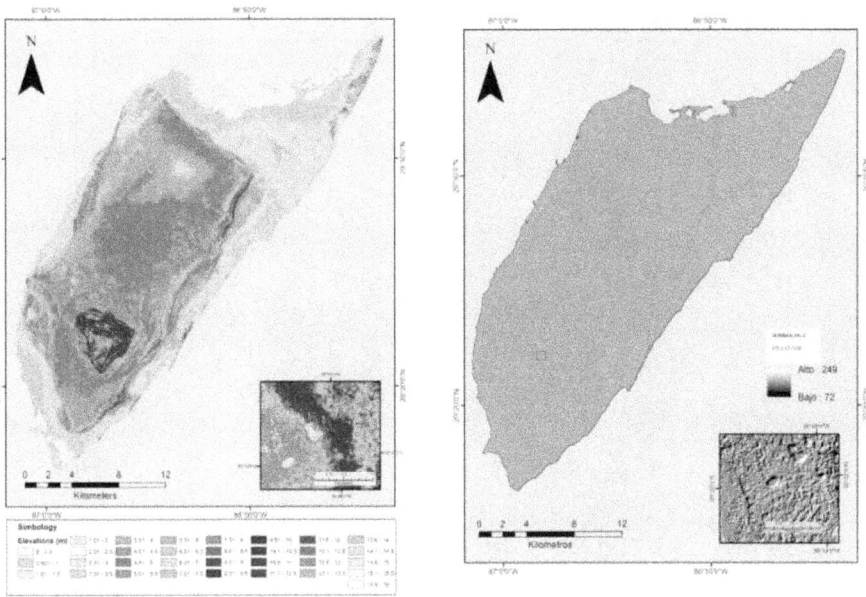

Figure 3. Elevation map with a vertical difference of 0.50 cm and terrain shading model (azimuth 227° and elevation 30°) with details about the known karst formations. Prepared by author based on LiDAR data [8].

was taken into account; groups that presented values in ascending order from the center to the shores were sought out, since this characteristic indicates that there are depressions. Furthermore, a visual interpretation was done in which the identification criteria was the geometric form of said groups. In this case, the semicircular form is associated with the dolines and the irregular forms with processes of the formation of uvalas. The dimensions are congruent with that which was identified in the shading model.

Finally, with the contrast of the slope (see **Figure 4**), where the association of the semicircular forms and forms with a gradient greater than 25° – corresponding, in the majority of cases, to the borders of the uvalas or collapse cenotes – the altimetric difference was up to 8 m of depth. The units with lower gradients (<25°) and altitude contrast (up to 1 m in height) are related to dissolution dolines.

In the records of dolines and uvalas in the region of study, the existence of 37 karstic forms are reported, all of which have been reported as points (with latitude and longitude coordinates), the majority of which measure less than 10 m in diameter [1, 17]; with the interpretation of the models derived from the LiDAR data, 109 forms were able to be identified, with their diameter and depth. The cartography of all of the sites is shown in **Figure 5**.

Furthermore, the density of dolines and uvalas in square kilometers is shown in **Figure 6**. The concentration can be explained by the fact that the area of the highest relief (>10 masl), which corresponds to a marine terrace in which the karstification processes are more evident than in the periphery as well as other morphological genesis (dunes, coastal mountain ranges, lagoons, and shoals), located along the coast and the north of the island, and whose formations can be buried or subterranean (such as caverns and grottoes).

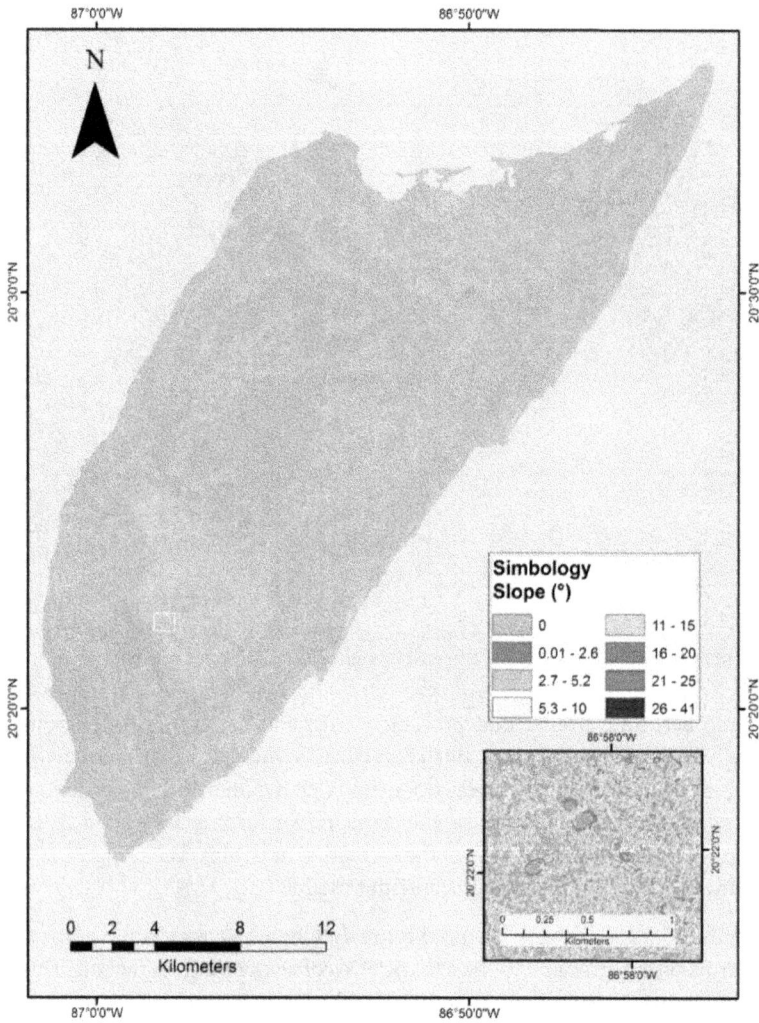

Figure 4. Slope map with six gradient categories. Prepared by author based on LiDAR data [8].

Although there is a new distribution model of dolines and uvalas, as well a proposal for density of karstic forms, it must be considered that in this study, the interpretation and identification of exokarstic forms follows a traditional process of relief form analysis [3, 5], based on the visual morphological differences (crests, slopes, surface, and background), known for their morphometric elements (height, slope, and depth).

The advantage of using LiDAR models, which are free for the public to access through the National Institute of Statistics, Geography and Informatics of Mexico, allows the processing of detailed data at a detailed horizontal scale (5 × 5 m) and at 0.05 cm of vertical height. The aforementioned was not possible until the year 2010, when LiDAR flights were carried out in

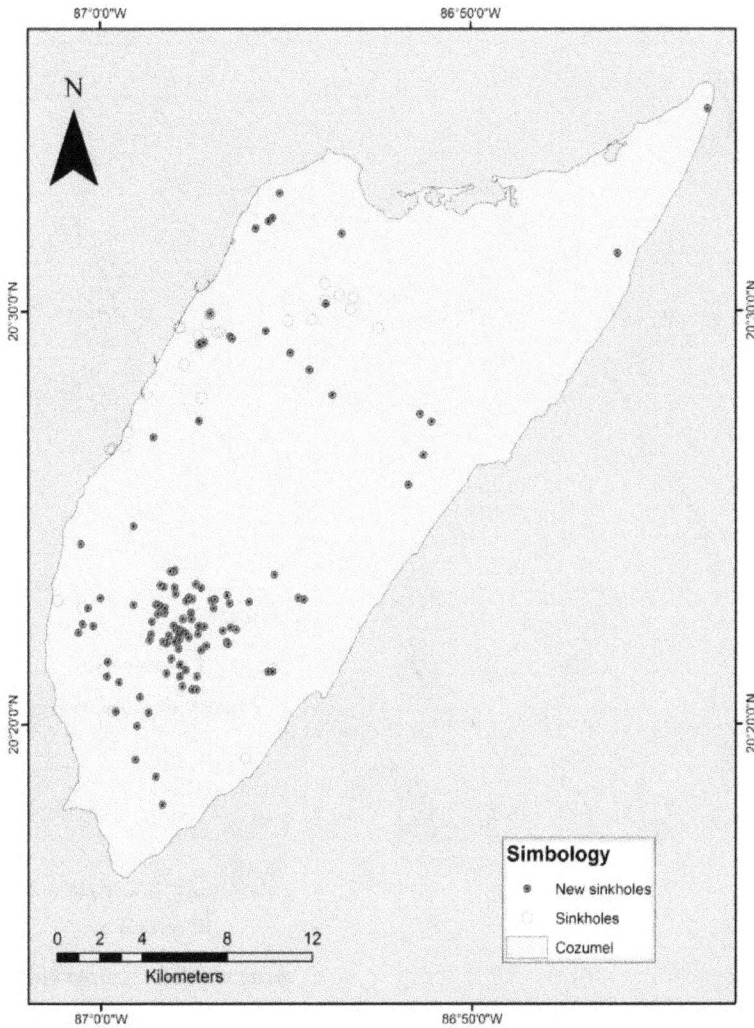

Figure 5. Map of point distribution of the karst forms in Cozumel Island. Prepared by author based on data reported [12–19] and this study.

this area of study, characterized by low altimetric contrast (<10 m) and by a base cartography at 1:50,000 scale. For this reason, it is not possible to identify karstic forms with dimensions less than 50 m, where the cost would be excessive, as pointed out by others authors [10, 14, 19].

New dolines and cenotes have been identified, along with the areas of greater density. However, an accurate characterization of each identified unit is needed to increase the under-standing of the types of dolines (dissolution, collapse, and suffusion) and uvalas (first, second or third generation), their relation with the creation (structural or climatic) and with the type and intensity of the process of karstification.

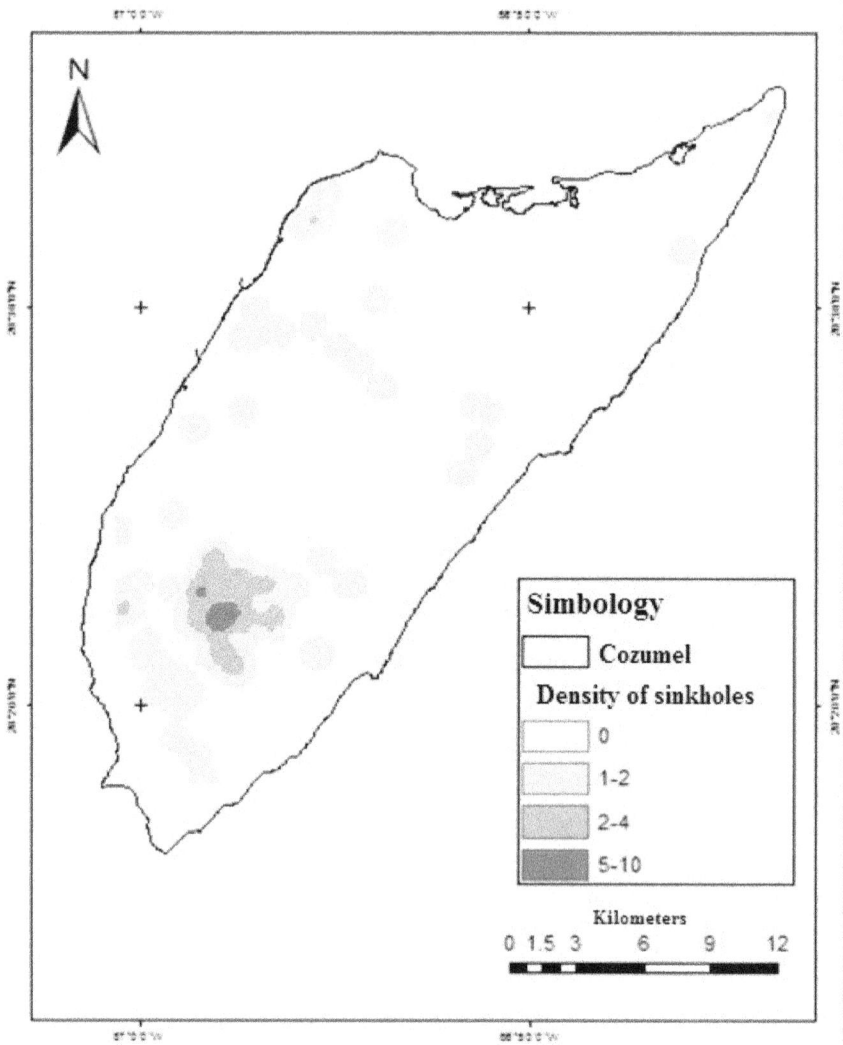

Figure 6. Density model of karstic forms by km² in Cozumel. Prepared by author based on the identification of dolines and uvalas using the LiDAR model [8].

4. Conclusions

The use of LiDAR data is not new to the study of karstic reliefs in tropical areas, where data has been manipulated by filters to highlight forms in the relief; among the most common are the hypsometric models, field studies, and slope.

In the area of study, there are no reports about the use of LiDAR technology for the recognition of dolines and uvalas, and consequently the use of LiDAR data is a proposal to identify and enrich the doline and uvala inventories in karst areas with little altimetry contrast. Even though in this study, new dolines and uvalas are reported, it lacks the complete verification and error estimation in the data interpretation and its concurrence with the reports in the literature.

The density map of dolines and uvalas have a resolution of 5 × 5 m, being a detailed-scale map which serves to orientate new searches for the calibration of LiDAR data and to be able to orientate its use in the entirety of the northeastern zone of the Yucatan Peninsula, where LiDAR data are open and free to access for scientific and academic study, with the potential for application to and studies about quaternary geology, the evolution of the landscape, the evidence of the coastal dynamic, and the transformation of the landscape in the area inhabited by the Maya of the Yucatan.

Acknowledgements

Thanks are due to the project Water and extreme hydrometeorological phenomenon in the Yucatan peninsula by CONACYT – REDESCLIM for their help. To the summer of scientific research for the help of student Angie Zapi and the trainee program at UQROO – University of Applied Sciences, Jena, Germany, for the help of student Christian Koch.

Author details

Oscar Frausto-Martínez[1,2*], Norma Angelica Zapi-Salazar[1,2] and Orlando Colin-Olivares[1,2]

*Address all correspondence to: ofrausto@uqroo.edu.mx

1 Laboratory of Observation and Spatial Research, University of Quintana Roo, Cozumel, Mexico

2 Sustainable Development Division, Universidad de Quintana Roo, Quintana Roo, Mexico

References

[1] Mejía-Ortíz G, Yáñez ML-M, Zarza-González E. Cenotes (anchialine caves) on Cozumel Island, Quintana Roo, México. Journal of Cave and Karst Studies. 2007;69(2):250-255

[2] Frausto O, Ihl T, Giese S, Cervantes A, Gutierrez M. Vulnerabilidad a la inundación en las formas exocarsticas del noreste de la península de Yucatán. Memorias del VI seminario latino – Americano de geografía Física, Universidad de Coimbra, maio de 2010. 2010. Available from: http://www.uc.pt/fluc/cegot/VISLAGF/actas/tema3/oscar

[3] Nayengandhi A, Brock J. Assessment of coastal vegetation habitats using LiDAR. In: Yang X, editor. Lecture Notes in Geoinformation and Cartography – Remote Sensing and

Geoespatial Technologies for Coastal Ecosystem Assessment and Management. Heidelberg, Germany: Springer Publications; 2008. pp. 365-389

[4] Mckean J, Nagel D, Tonina D, Bailey P, Wright C, Bohn C, Nayengandhi A. Remote sensing of channels and riparian zones with a narrow – beam aquatic – terrestrial LiDAR. Remote Sensing. 2009;**1**(4):1065-1096

[5] Zhu Y, Taylor TP, Currens JC, Crawford MM. Improved karst sinkhole mapping in Kentucky using LiDAR techniques: A pilot study in Floyds Fork Watershed. Journal of Cave and Karst Studies;**76**(3):207-216. DOI: 10.4311/2013ES0135

[6] Brock C, Purkis S. The emerging role of LiDAR remote sensing in Coastal research and resource management. Journal of Coastal Research SI. 2009;**53**:1-5

[7] Bayer JM, Scheid JL, editors. PNAMP Special Publication: Remote Sensing Applications for Aquatic Resource Monitoring, Pacific Northwest Aquatic Monitoring Partnership. Washington: Cook; 2009. 100 p

[8] Aguilar FB, Mendoza ME, Frausto O, Ihl T. Density of karst depressions in Yucatán state, Mexico. Journal of Cave and Karst Studies. 2016;**78**(2):51-60. DOI: 10.4311/2015ES0124

[9] Delgado-Carranza C, Bautista F, Orellana-Lanza R, Reyes-Hernández H. Classification and Agroclimatic Zoning Using the Relationship between Precipitation and Evapotranspiration in the State of Yucatán. Vol. 75. Mexico: Investigaciones Geográficas, Boletín del Instituto de Geografía, UNAM; 2011. pp. 51-60. DOI: 10.14350/rig.29795

[10] Wurl J, Giese S. Ground water quality research on Cozumel Island, State of Quintana Roo, Mexico. In: Frausto Martínez O, editor. Desarrollo Sustentable: Turismo, costas y educación. México: Universidad de Quintana Roo; 2005. pp. 171-176

[11] INEGI. Modelo digital de elevación LiDAR. México: Instituto Nacional de Estadística, Geografía e Informática; 2011. Available from: http://www.inegi.org.mx/geo/contenidos/datosrelieve/continental/presentacion.aspx

[12] Orellana R, Nava F, Espadas C. El Clima de Cozumel y la Riviera Maya. In: Mejía-Ortíz LM, editor. Biodiversidad acuática de la isla de Cozumel. México, D. F.: Universidad de Quintana Roo – Plaza y Valdés; 2007

[13] Fragoso-Servón P, Pereira A, Frausto O, Bautista F. Geodiversity of a tropical karst zone in South-East Mexico. In: Andreo B, Carrasco F, Durán J, Jiménez P, LaMoreaux J, editors. Hydrogeological and Environmental Investigations in Karst Systems. Environmental Earth Sciences. Vol. 1. Berlin, Heidelberg: Springer; 2015

[14] Coronado-Álvarez L, Gutiérrez-Aguirre M, Cervantes-Martínez A. Water quality in wells from Cozumel island, Mexico. Tropical and Subtropical Agroecosystems. 2011;**13**(2): 233-241

[15] Fragoso-Servón P, Pereira Corona A, Bautista Zúñiga F, Gonzalo de Jesús Zapata B. Digital Soil Map of Quintana Roo, Mexico. Journal of Maps. 2017;**13**(2):449-456. DOI: 10.1080/17445647.2017.1328317

[16] Cervantes-Martínez A, Gutiérrez-Aguirre M, Álvarez-Legorreta T. Indicadores de calidad del agua en lagunas insulares costeras con influencia turística: Cozumel e Isla Mujeres. Quintana Roo, México: Teoría y Praxis; 2015. pp. 60-83

[17] Fragoso-Servón P, Bautista F, Frausto O, Pereira A. Caracterización de las depresiones kársticas (forma, tamaño y densidad) a escala 1:50,000 y sus tipos de inundación en el Estado de Quintana Roo, México. Revista Mexicana de Ciencias Geológicas. 2014;31(1): 127-137

[18] Yañez G. Sinkhole Database of Cozumel. Cozumel, Mexico: Mayab AC; 2015. Available from: http://speleomayab.blogspot.mx/

[19] Frausto-Martínez O, Ihl T, Rojas López J. Risk Atlas of Cozumel Island, Mexico. Teoría y Praxis [en linea]. 2016 (Octubre-Sin mes). Availble from: http://www.redalyc.org/articulo.oa?id=456147940005

[20] Weishampel J, Hightower J, Chase A, Chase D, Patrick R. Detection and morphologic analysis of potential below-canopy cave openings in the karst landscape around the Maya polity of Caracol using airborne LiDAR. Journal of Cave and Karst Studies. 2011;73: 187-196. DOI: 10.4311/2010EX0179R1

[21] Gesch DB. Analysis of LiDAR elevation data for improved identification and delineation of lands vulnerable to sea-level rise. Journal of Coastal Research: Special Issue. 2009;53:49-58

Lessons Learned from the Establishments of the First Hydrographic Surveying Program in the Middle East

Rifaat Abdalla and Salim Al-Harbi

Additional information is available at the end of the chapter

http://dx.doi.org/10.5772/intechopen.82527

Abstract

The fast pace of technology development and voluntary adoption of international standards requires interdisciplinary and skill-based education. This chapter presents an approach for the development of an interdisciplinary, internationally recognized geomatics program, at King Abdulaziz University (KAU), using a multilevel approach that combined the international guidelines with the local stakeholders' needs being in line with the global demand for professionals in this field. The methodology of this study consisted of interviews with subject matter experts (SMEs), students survey and operational analysis, and observation was used to analyze the program challenges and opportunities. Results obtained showed that the transferability of the approach adopted in this research, along with the commonality of lessons learned from the process, contributes to faster execution for similar programs in various parts of the world. The program was successful to secure to international recognition within 10 years of its inception. The quality of learning outcomes supported by the high employability of graduates was among the key socioeconomic impact of the program.

Keywords: IHO, geomatics, hydrography, curriculum design, learning outcomes, system model, action research, systems thinking

1. Introduction

Geomatics is the new discipline that integrates the tasks of gathering, storing, processing, modeling, analyzing, and delivering spatially referenced data or location information [1]. The spatial technologies represent the core of geomatics and help determine the location and identifying the bathymetry of water bodies. The broad application of geomatics technologies in marine and oceanography applications has allowed geomatics to integrate all the elements of spatial sciences and remote sensing along with measurements in a unique discipline

IntechOpen

known as hydrography. Hydrography is defined as the science of mapping and charting the depths of water, whether it is seas, lakes, rivers, or oceans. There are many supporting fields to hydrography including coastal zone management, nautical charting, the safety of navigation, ocean mapping, marine resource exploration, maritime boundary delimitation, protection of the marine environment, marine science, and naval activities of defense. The importance that hydrography gained in the recent years is because of the collective efforts ongoing and the justified need for more additional work on further exploring and exploiting natural resources available in marine environments. The IHO estimates that at least 50% of the world's coastal waters are unsurveyed. The polar regions, the South West Pacific, West Africa and the Caribbean are about 10% surveyed. Moreover, in those areas where studies do exist, many are so old or of such a quality that they cannot support the modern requirements. A direct influence on measurements and observations related to climate and climate change makes hydrography a key enabler to the sustainable development of the seas and the best management and governance of the ocean sustainability and resources.

Until about 20 years ago, the traditional components of geomatics, namely: photogrammetry, cartography, remote sensing, and surveying were all independent, and each had its distinct identity [2]. However, geomatics development is directly attributed to the advances in computer science [3]. To this end, there is a debate about whether geomatics is an evolution of the traditional surveying engineering or natural development in the field of earth science. Regardless of the origin of geomatics, the fundamental fact is that it is not possible for a single person, i.e., surveyor or computer scientist to provide a complete solution at the required knowledge depth necessary [2]. Geomatics provides collaborative solutions have a broad range of applications [4], which makes it a unique discipline. Applications such as spatial database design and management, environmental engineering and climate change modeling, oceanography, forestry, geology, geophysics, civil engineering, and biology have made geomatics as a hub for subspecialties of high professional and economic interest [5, 6]. **Figure 1** show the interdisciplinary science of Marine Geomatics.

The objectives of this chapter address the challenges and highlight the opportunities that arise from developing marine geomatics program in the Kingdom of Saudi Arabia, as an international model of collaboration for similar international efforts. The study was planned to achieve three primary goals, specifically:

1. The utilization of maritime resources is critical to ensuring the economic well-being of many economies. For the Gulf Cooperation Council (GCC) region, the region requires increasing academician awareness with challenges and opportunities as a result of developing a new marine geomatics program with regional specifics of geographic and socio-cultural constraints, as well as with international impact and contribution.

2. For local development to occur, it is important to invest in the local establishment of "approach-based engineering education," despite the high costs and other challenges, such as meeting multiple stakeholders' requirements in skill-based education. Such development requires an increased level of the knowledge of the role of the local and international stakeholders' contribution to the development of a new marine geomatics program.

3. Evaluation of the program outcomes and accomplishments in 10 years since its inception is necessary to ensure the establishment of the involved knowledge and its generational transmission.

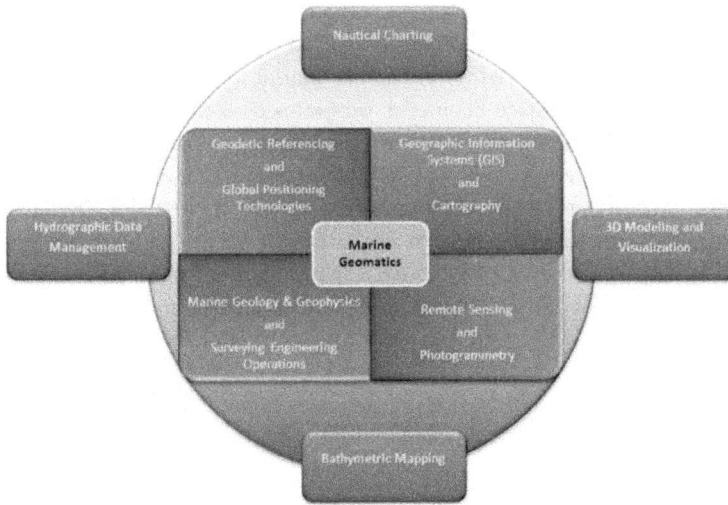

Figure 1. Marine geomatics as interdisciplinary science.

The approach investigated the need and challenges faced to supply the job market with highly trained professionals that are internationally certified to carry out their job, regardless of their regional boundaries. This study discusses issues related to challenges and opportunities for geomatics education and showcases the development of the new hydrographic surveying program at KAU, in the heart of a large metropolitan port city of Jeddah, Saudi Arabia. The study highlights particular issues as a model for geomatics education in similar parts of the world.

In the following sections, the paper sets the foundation for justifying the need for hydrographic surveying program, examines students' performance and satisfaction through survey analysis, and explores current challenges through subject matter experts (SMEs) interviews. It outlines the development of the curricula and introduces the challenges faced by the program and the opportunities that await the new curriculum as it relates to KAU. The last section of the paper draws conclusions within the context of challenges and opportunities associated with the development and evaluation of new specialized geomatics programs.

2. Literature review

2.1. Geomatics education

The significance of approach-based engineering education research has been documented by many researchers, including [7]. Several interdisciplinary subdomain research methods have received substantial attention in education research due to their emergence to form multidisciplinary approach. Case and Light and Cousin [8, 9] indicated that attribution of understanding and implementation of tools as procedures of inquiry in engineering research methods is to the way of their application as tools and context interpreted. The framework formed by the instruments and methods is known as "methodology."

The focus in engineering education research methods that focus on the process was the focus of the work of many researchers including [8, 10]. Today, many advanced economic systems are challenged to rely conclusively on and utilize marine resources. Additional importance linked to having effective marine geomatics programs as a priority for many parts of the world. Agrafiotis and Koumoutsos [11] defined the term education as the process of ensuring the development of knowledge, the formulation, and adaptation of the acquired knowledge to the collective memory and its processing, which contributes to making this process ongoing between generations. Many researchers have discussed the need for geomatics education including [2, 12–14]. Regionally, the need for a specialized geomatics education in the Gulf Cooperation Council (GCC) region comes from the fact that there is a growing population versus natural resources in the area that are underwater, and there are ongoing efforts for exploitation. Another factor is that the GCC countries are in a peninsula surrounded by water, between three of the major water bodies i.e., the red sea, the Indian Ocean, and the Arabian Gulf, which makes safe navigation of increased importance to the region. The fast development of maritime infrastructure and transportation and the large fleet of ships and oil tankers pose a new concern for programs related to marine transportation and safety.

The high cost of establishing efficient maritime education infrastructure, as well as the hard nature of offshore training that students and professionals in maritime industry require, adds more challenge to having enough teaching and training facilities. The outcome of that is a less human resource that can share and transfer the knowledge to the future. It is a global situation faced by a very limited number of specialized marine geomatics programs worldwide. There are many unclear boundaries for the connection between marine geomatics as engineering discipline at its link to many marine sciences including marine geology, marine applied physical oceanography and environmental sciences. The lessons learned from developing technology-based education are attributed to the IT infrastructure itself, as well as to adequately addressing the environmental consideration by the international standards, as discussed by Agrafiotis and Koumoutsos [11]. The need for technologically supported education is on the rise as all educational activities today are helping the process of growing economies, and they depend on technology to a far extent.

The growth of geomatics sector in Saudi Arabia is expanding, along with the process of outgrowing the challenges associated with technology adoption and utilization. This increase makes a golden opportunity for local training of professionals in the field hydrography, according to the international standards of competence laid by the International Hydrographic Organization (IHO). However, despite the observed development in geomatics, it remains limited compared to the western world [15]. Konecny [5] believes that due to the many factors supporting effective adoption of Geomatics Technologies, the need for formal education need is also growing in this advanced technological era. This realization is most pertinent since in an actual sense, development is never achieved nor can it be sustained from outside the developing country [16]. Local professionals play a strategic role in the socioeconomic development of their communities [17]. This provides added value justification for the need of having a state-of-the-art education that combines technology capabilities with local needs. In many developing countries, there is the absence of effective local participation and involvement in strategic planning, formulation, program identification, design, and implementation [16].

Therefore, the spreading of technology-based programs that provides geomatics solutions has to go along with a detailed analysis of the local and regional situation [6, 18].

The scope of skills and expertise required to form the link between higher education institutions of today, whether in form college or university education. The global demand for geomatics professionals and hydrographers worldwide reinforces the importance of having marine geomatics program in the Kingdom of Saudi Arabia, mainly due to the following:

- some of the geomatics-related works in the region are still in the realm of standard (traditional) practice,

- the rapid development in computation and adoption of digital forms of data processing and conversion,

- the clear link between socioeconomically developed communities, sustainability of corporations and government agencies, and the realized need for proper economy drivers that help with wealth data collection and handling,

- the continuing advances in environmental protection, sustainable development, and natural resources conservation through the adoption of advanced Geomatics Technologies that require highly qualified and highly trained personnel.

Marine geomatics is an interdisciplinary applied science that is based on the foundation of Geodesy and Land Surveying. The program is designed on the pillar of four core domains: (1) geodesy and positioning, (2) land surveying and estimation, (3) remote sensing and photogrammetry, (4) GIS and cartography, (5) oceanography and marine environment, and (6) marine geology and geophysics. The components from 1 to 4 are standard in any geomatics or land surveying program. However, five and six are unique in this program and many other hydrography/marine geomatics programs worldwide. The strength of marine geomatics program at King Abdulaziz University is directly attributed to the stakeholders' interest in having professionals in this field supported by a strategic partnership with Saudi Aramco, the largest oil company in the world. Our program is designed to hosting top talented students and to provide a high-quality education according to the international standards of the International Hydrographic Organization (IHO). The international recognition of the program by the IHO has contributed to the excellence and strength of the program.

2.2. Specific needs for the establishment of marine geomatics program

As suggested by Refs. [19, 20], the technology advancement has allowed for developing a new program to cross the interdisciplinary horizon of all sectors of Information Technology. Marine geomatics is one of these interdisciplinary programs that is critically required due to existing gaps in geomatics education in the region and absence of the hydrography-related, skill-based program. This makes the initiative of KAU unique, not only in the Kingdom but also in the region. The fast growing pace of geomatics as interdisciplinary skill-based education in North America has triggered the global need for such education, the relatively rapid pace growing in supporting fields [5, 21].

More specifically, the need for engineering-based multidisciplinary marine geomatics education is increasing day by day. There is an expanding global change in economic dependence in soul natural resources products, such as oil, and the growing concern of utilization diverse financial resources that can deal with an array environmental challenges [22]. All these in addition to the growing technological advancements have maximized the need for skill-based education for professionals, specifically for Saudi Arabia and the GCC region. As indicated by Melezinek [23], skill-based technologically obsessed education became necessary as the application of knowledge became as important as it is a pursuit. This type of education needs to be supported by a contemporary approach to providing advanced professional education. The required efficient and progressive decision-making process has helped with shaping and advancing geomatics education [11]. Today, it is not only the academic community that is concerned with the issues of providing adequate advanced skill-based education, but is also the stakeholders, who are more concerned to have knowledgeable and skilled professionals that can support their communities [24]. The academic education and professional training integration to pro-vide skill-based education have become a need, rather than a complementary resource, more particularly to developing countries with growing economies and depleting resources [25]. It partially addresses the need raised by many researchers including [11] who illustrated the need in keeping up with the rapidly developing technologies through active education systems to provide advanced knowledge and to enhance the contribution to the development of vibrant communities that support stakeholders' objectives and job market trends. This justification is according to the growing need for effective educational systems that are capable of providing advanced training that is keeping with the rapid pace of technology development [12, 14, 26].

3. Methodology

This research is considered as action research, as it summarizes efforts being carried out in the process of developing a new educational program. It introduces a new approach that inte-grates the environmental socioeconomic considerations with the international requirements for providing advanced skill-based education and training.

It provides direct input to methodologies and mechanisms that are currently embraced glob-ally to improve the process of creating similar programs. The relevance of this approach taken is not limited to the region or country where the development occurred. It can be adopted worldwide with minimal consideration to the socioeconomic factors that might influence suc-cessful implementation of such a program. In geomatics, there are many efforts presented by many. The research adopts different levels of analysis and observations, based on the approach proposed by Virkki-Hatakka et al. [27]. **Figure 2** shows a summary of the research approach used in this study. The reason behind this systematic approach was to evaluate challenges and opportunities for marine geomatics programs using local case study.

The research is intended to reach to results and develop conclusions based on critical evalua-tion of the approach adopted in the development of the academic program. These assessments are important because they provide heuristic evaluation and key observation for the successes and failures of the program. The first stage of the methodology involved SMEs' interviews,

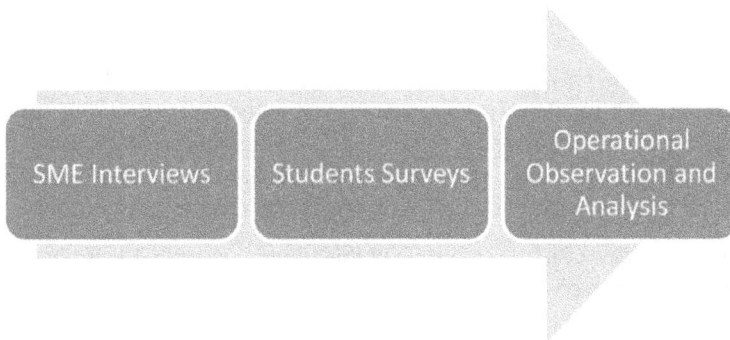

Figure 2. The approach for research.

where two subject matter experts, from those who participated in the evaluation of the program, have analyzed the strides and failures of the program, based on the standard assessment procedure that has focused on determining and listing the challenges and opportunities that are faced by the program. The levels adopted are as follows: (A) critical evaluation of the strategic plan adopted in the developed program, based on qualitative input from different stakeholders and academic administrators; (B) analysis of the policies and procedures used during implementation of the plan while developing the program; (C) observation, evaluation, and self-evaluation of the program performance and development over the study years; and (D) reflection on the outcomes based on the steps from A to C, where they were assessed based on the contribution of the program to addressing the stakeholders' needs for local professionals according to the international standards. The second stage of the research has focused on analyzing the information obtained from the annual students survey completed as part of the quality monitoring assessment, under the University Vice-President Office for development, where students complete the survey. The methodology also included analysis of socioeconomic indicators for the success of the program at a regional level. Critical analysis of the impediment factors for further improvement of local and regional considerations is associated with the establishment of the program.

4. Case of KAU hydrographic surveying program

The hydrographic surveying department aims to provide graduates with basic and advanced knowledge of hydrography, hydrographic data collection, management, and presentation. Also, it provides knowledge of data analysis to determine depths and locations and to create hydrographic/nautical charts, as well as navigational and environmental protection. It has started in 2004 with two full-time faculty members and four collaborating faculties, to reach to eight full-time faculty members, two engineers, and four faculty members under development. The justified need has led KAU to initiate a program that builds the region capability for hydrographic surveying cadre. It also supports the ongoing development and exploitation of maritime resources. Since its inception and for 7 years, the program was under the Department of Maritime Studies at King Abdulaziz University and was hosted by the Faculty of Marine Science of KAU. In 2011 the announcement of the establishment of a new Faculty of Maritime studies was announced by approval from the higher authorities in Saudi Arabia (Royal Decree). The new Faculty of Maritime Studies hosts four departments, which are the Department of Hydrographic Surveying, the Department of Nautical Science, The Department of Ports and Maritime Transport, and the Department of Marine Engineering. The department graduates are 20 in total, with the first batch graduated in 2008.

The department facilities consist of the state-of-the-art equipment and instruments, including the "University Hydrography" survey vessel. The department services are among the most comprehensive regarding both technical and academic aspects. In April 2008, the construction was completed in the full five-storey maritime studies building, at KAU Campus Al-Morgan, 40 km north of the main campus in Jeddah. This building location is in the creek of Abhor and the Red Sea. A new 16-meter long hydrographic survey vessel was acquired in 2010 to provide hydrographic surveying students with all what they need to practice hydrographic surveying in the field. The vessel utilizes the state-of-the-art hydrographic surveying equipment and processing packages. It allows up to eight students and four crew members to conduct field survey for up to 8 days onshore using onboard data processing and transmission to different receivers.

4.1. Academic requirements

The program is unique in the Gulf Cooperation Council (GCC) states; it is designed to address the gaps in educating professionals in the region, as no other institution offers a similar program or professional certification. The International Maritime Organization (IHO) [28] guidelines provided direct input in the development of the program. The program has focused on addressing far beyond the core competencies needed for hydrographers CAT A certification. The information provided by the local stakeholders [29], including the Saudi Aramco, Ministry of Transportation and National Surveying Authority, provided a regional benchmark for the skills and competencies required. The stakeholders showed a wide range acceptance and support of the idea of providing a professional license (CAT A) along with

Figure 3. KAU "hydrographia1" training vessel.

B.Sc. in hydrographic surveying. **Figure 3** showing the University Hydrographia 1 survey vessel which is used a field data collection and survey planning training lab.

The B.Sc. in hydrographic surveying program is designed for 131 credit hours, ideally, over 4 years of study [30]. **Table 1** is shows the program structure. The first year of the program covers the university requirements, as the education system in Saudi Arabia requires university students to take a faculty-based previous year, in which students will study 14 credits of common foundation courses in natural sciences as outlined in **Table 2**. As discussed by Rashid

No.	Course name	Course code	Units
1	Mathematics for Maritime Studies	MATH 201	4
2	Linear Algebra I	MATH 241	3
3	Physics 204	PHYS 204	4
4	Physics (practical)	PHYS 281	1
5	Introduction to Computer Science	CPIT 201	3
6	Maritime Communication	MSN 243	2
7	Marine Meteorology	MSN 245	2
8	Fundamentals of Surveying	MSS 220	3
9	Technical Communication	MSS 221	2
10	Fundamentals of Nautical Science	MSS 310	3
11	Underwater Positioning Systems	MSS 311	3
12	Estimation and Uncertainty Management	MSS 312	4
13	Geodesy and Map Projections	MSS 313	4
14	Water Level Measurements and Prediction	MSS 320	3
15	Surface Positioning Systems	MSS 321	3
16	Underwater Imaging and Mapping I	MSS 322	4
17	GIS and Bathymetric Data Management	MSS 323	3
18	Marine Law and Policy	MSS 410	1
19	Underwater Imaging and Mapping II	MSS 411	4
20	Remote Sensing for Marine Applications	MSS 413	3
21	Marine Geology for Hydrographers	MSS 414	3
22	Nautical Charting	MSS 420	3
23	Hydrographic Data Management and Presentation	MSS 421	3
24	Degree Project	MSS 429	3
	Total		71

Table 1. Hydrographic surveying core courses.

1	The program structure (including the knowledge and skills to be acquired) was evident to me
2	The things I had to do to succeed in the program including courses assessment tasks and criteria for assessment were made clear to me
3	Sources of help for me during the program including faculty office hours and reference material were made clear to me
4	The conduct of the courses and the things I was asked to do were consistent with the course outline
5	My instructor(s) were fully committed to the delivery of the courses (e.g., classes started on time, instructor always present, materials well prepared, etc.)
6	My instructor(s) had a thorough knowledge of the content of the courses
7	My instructor(s) were available during office hours to help me
8	My instructor(s) were enthusiastic about what they were teaching
9	My instructor(s) cared about my progress and were helpful to me
10	Courses materials were of up to date and useful (texts, handouts, references, etc.)
11	The resources I needed (textbooks, library, computers, etc.) were available when I needed them
12	In this program, efficient use was made of technology to support my learning
13	I was encouraged to ask questions and develop my ideas
14	I was inspired to do my best work
15	The things I had to do (class activities, assignments, laboratories) were helpful for developing the knowledge and skills
16	The amount of work I had to do was reasonable for the program credit hours allocated
17	Marks for assignments and tests were given to me within a reasonable time
18	Grading of my tests and assignments was fair and reasonable
19	The links between the courses and throughout the program were made clear to me
20	What I learned is important and will be useful to me
21	I had improved my ability to think and solve problems rather than just memorize the information
22	I was able to develop my skills in working as a member of a team
23	I had improved my ability to communicate effectively
24	Overall, I was satisfied with the quality learning experience

Table 2. Students survey questions.

and Tasadduq [31], some programs face the challenge in curriculum design as it relates to the rapid evolution of technology, which was a major consideration in exploring the options for setting this academic program to meet the IHO professional standards.

The advanced curriculum addresses the university requirements, the faculty requirements and the program core courses and elective courses requirements as set by the King Abdulaziz University, Academic Policy. A total of 131 credits covers some 14 credit hours, divided into four courses at the preparative preceding year, according to the Saudi higher education system. At this prior year, the students learn mathematics and physics extensively. Another 26 credit hours represent additional courses according to the requirements of KAU. In this, 26 credits of compulsory university requirement courses include Arabic and English languages, Islamic culture, and Humanities. The university requirements are also supported by 10 credits of the faculty requirements, in which the faculty needs students to study mandatory faculty courses, namely marine environment, statistics, and oceanography. A total of 50 credits that constitute the university requirements, as well as the faculty requirements are a must prior to starting the core courses of the Hydorgprahic Surveying specialization. Out of 81 credits, 69 credits represent the mandatory courses and 8 credits represent the elective courses that students can choose from a vast number of classes. The last four credits are for the necessary training as the IHO requires it.

English is the language of study in the program. Students must complete intensive four courses in English during their first 2 years. Ideally, before the start of the specialization courses by the beginning of the third year, students must obtain an equivalent to 200 TOEFL score, to begin their third year. KAU English Language Centre offers these courses.

The core program credits are 81 credit hours and provide students with all the skills needed to become a professional hydrographer; starting at the seventh term, students begin to take specialty courses. Students must mainly go through the process of applying the scientific method in solving particular research issue and report that in a sound acceptable precise way as a graduation project. The core hydrographic curriculum covers four specialty areas, which are: Hydrographic Surveying courses, where students will learn about the methods and procedures of hydrographic surveying, supported by the Science and Engineering behind that. In the core courses students are introduced to to tide measurement, echosounder equipment measuments through singlebeam and multibeam echosounding, sidescan sonar, offshore geophysical surveying. The second specialty area covers positioning and navigation, including terrestrial and satellite positioning and altitude systems. This component will provide the student not only with the basic foundations needed to meet the requirement as Category "A" hydrographer but also to address the academic requirements for getting a B.Sc. degree. The third specialty area covers geodesy and estimation, where students will learn about datum and coordinate systems, map projections, maritime boundary delineation, estimation and filtering, and uncertainty management. This component will provide the student not only with the basic foundations needed to meet the requirement as Category "A" hydrographer but also to address the academic requirements for getting a B.Sc. degree. The fourth specialty area is nautical chart production, where students will learn about Geospatial Information Systems (GIS), data management, remote sensing, and photogrammetry. This component will provide the student not only with the basic foundations needed to meet the requirement as Category "A" hydrographer but also to address the academic requirements for getting a B.Sc. degree. **Table 1** provides an overall view of the components of the hydrographic surveying program and their term distribution, including prerequisite and elective courses.

Students can choose to take 8 credit hours of elective courses to add additional 8 credits in order to graduate. The courses students can choose from are Technical Communications,

Marine Geology for Hydrographers, Offshore Geophysical Surveying and special topics in hydrography.

The field training provides the required field skills as outlined by the International Hydrographic Organization (IHO) guidelines [28]. Over two terms of 6 weeks on the board of a hydrographic survey vessel, students gain the required practical knowledge needed to work with different data acquisition systems. The focus of the core skills that students get while on training covers field calibration of single beam/multibeam echo sounder, multibeam system patch test, and reference systems. The students are also involved in the survey design and planning, hydrographic surveying specifications, and types of hydrographic surveys.

4.2. International certification and local accreditation requirements

The International Hydrographic Organization (IHO) was formed in the year (1921) by some member states, with principal objective to ensure that all the world's seas, oceans, and navigable waters are surveyed and charted, to provide safe navigation for mariners. Its vision is to act as the sole authority worldwide to provide governance and guide for global hydrographic activities. The IHO is a United Nations observer organization. Its mission is "to create a global environment in which States provide adequate and timely hydrographic data, products and services and ensure their widest possible use" [28]. Some 85 coastal states are engaged as members of the IHO and work on promoting and advancing maritime safety, including the protection and sustainability of the marine environment. The international board handles the accreditation process in the IHO on Standards of Competence for Hydrographic Surveyors and Nautical Cartographers (IBSC). It regulates and accredits academic programs and departments that provide certification for professional hydrographers and nautical cartographers. The competency standards are according to Standard 5 revision 11 of the IHO regulations [28].

The Department of Hydrographic Surveying offers a Bachelor of Science degree (B.Sc.) in hydrographic surveying. The program was designed to meet the requirements of the International Hydrographic Organization (IHO) as Category A (CAT-A), according to the latest revised edition of the Standards of Competence for Hydrographers guidelines [28]. The program was recognized in April 2013 as IHO accredited program for (CAT A) professional certification. IHO accreditation provides worldwide recognition for graduates in their level of competence to perform advanced hydrographic surveying skills globally, regardless of their service region. The process of accrediting the Department of Hydrographic Surveying by the IHO was started in 2007 when the Saudi General Commission invited the president of the IHO for survey (GCS). The preparation of the department profile for submission to the IHO involved discussions among the department, the GCS, and the IHO; this has initiated the communication with the IHO. Later, in 2007, the Saudi authorities organized an international workshop in Capacity Building, hosted by local authorities and convened in the City of Jeddah. In December 2012, the Department of Hydrographic Surveying presented its portfolio at the annual meeting of the IBSC in April 2013, where the recognition of the program as Category "A" accredited institution, the highest recognition in the IHO scheme of two categories was granted.

The international interest expressed by the International Hydrographic Organization (IHO) was evident. The Capacity Building administration of the IHO was working hard to expand the presence of IHO- certified hydrographers in many regions of the world. A high delegation from the IHO has visited Saudi Arabia several times in the past 10 years. A meeting with the president of the IHO was held during his visit to Saudi Arabia to attend activities related to the GCS. A second meeting with the IHO representatives took place during the IBSC visit to attend a regional event in 2014. These two sessions with the administration revealed keen interest for support from the IHO, for the establishment of an international program at King Abdulaziz University.

The hydrographic Surveying Department has maintained a strong international support and collaboration with many organization including the Interdisciplinary Centre for the Development of Ocean Mapping (CIDCO) in Canada in hosting student training programs. CIDCO is a marine geomatics R&D organization that hosts IHO recognized program. ENSTA Bretagne in France has also provided support to the program by delivering training for students in 2012. ENSTA Bretagne is a French national graduate engineering institute with reputed contribution to hydrography. A delegation of the chapter of the UK Hydrographic Society in UAE has visited the program and showed interest in sponsoring professional talk series in UAE, where representatives of the program can participate and share insight with practitioners into the domain of hydrography.

The Saudi Council of Engineers (SCE) has approved the membership of the graduates of hydrographic surveying program. The objectives of SCE are to promote the engineering profession. The council exercises many roles to do whatever may be necessary to develop and upgrade its standards, however, currently the membership of Hydrographic Surveying graduates to the SCE is under review [32]. The mandate of the SCE stipulates that it determines the suitability of the program in terms of accreditation requirements. Till the year 2015, the graduates of the program are recognized as individual members of the SCE according to the SCE regulations.

4.3. Stakeholders' requirements

The support that hydrographic surveying program gets goes far beyond the university top academic administration. The initiation of the program was completely backed up by an advisory board of government and industry. This support secured that the program meets the need of local and regional employers, and provides international standards for training and education. Specific support was evident from the Saudi Aramco, Saudi Ports Authority, Ministry of Transportation, and the Saudi Military Survey Department. Some mutual visits and consultations with the Saudi General Commission for Survey (GCS), Saudi Navy, the Saudi Geological Survey, and the Saudi Coast Guard helped with shaping their operational requirements. The success of the program in securing support from government stakeholders, as well as private stakeholders represented by industry members from Saudi Arabia, provided added value to the program design. Regional support from relevant stakeholders from the United Arab Emirates, who expressed interest in attracting students to work with them on various projects, represented another evidence of the program success. The foundations for collaboration and support have emerged in some mutual agreements and MOUs. All graduates of the program in 2014 were hired by GCS as hydrographers and resumed their duties with positive feedback from their managers.

The program was successful to form industry advisory committee that brings all relevant stakeholders. The committee was successful in providing requirements and considerations for future employees, in the field of hydrographic surveying. Special extra meetings with Saudi Aramco, the leading international oil company and the General Survey Commission (GCS), in addition to leading private sector enterprises, provided some value insights into the program development.

5. Discussion of the findings

The data used for this study came from all three stages of the methodology as outlined in the methods section. Over the past 10 years, the program has granted B.Sc. degree to a total number of 60 students. The majority was graduated before the recognition of the IHO was given. In 2016, the first batch of the program with 10 students has graduated with the IHO recognition as Category "A," which makes this as a milestone in the application development by providing the graduates with international certification along with B.Sc. in Hydrographic Surveying. Out of 50 students, a total number of graduates from the program over the study period, the survey covered 55 students for their feedback on their evaluation of the program and whether the program has met their expectations regarding quality of education needed for the job, or in supporting their future career objective. The interview questions covered the knowledge requirements according to the Saudi National Commission on Academic Accreditation and Assessment (NCAAA) routine assessment for academic programs.

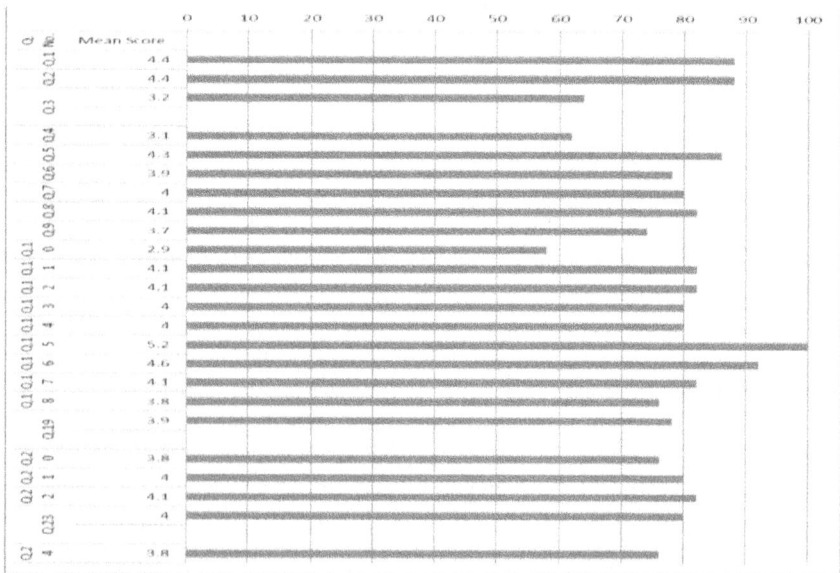

Figure 4. Mean response score % for the survey questions.

The questions of the Interview has focused on different themes, the first topic was general questions, the second theme was focusing on what happened during the program, and the third issue was the overall process evaluation of the program, and the last subject was an assessment of the overall satisfaction with the program. The survey questions are shown in **Table 2**.

The results of the student's survey as shown below have highlighted the level of satisfaction a large number of students have expressed in the program were rated on a scale of 1–5 where five were extremely satisfied, and one is extremely dissatisfied. The results as seen in **Figure 4** have all presented a good level of satisfaction except in 10, where some students have raised a flag on the course reference. This was before securing the IHO recognition, where the course material used was not very well structured. However, the matter has been improved in the following years. These results are cumulative results for the entire period of the study. The (n) is equal to 35 from 10 years time.

The graduate's exit survey has utilized the NCAAA standard forms, and it was used to highlight the themes of help and support for learning, resources available for learners, and the evolution of student and the overall evaluation of the program. The number of students who completed the exit survey was 35 students; they answered some questions dealing with their overall experience in the program, and it was satisfactory and supported 100% employment rate within 6 months of graduation.

5.1. Analysis of challenges

The SMEs interviews have focused on analyzing challenges faced by the program, which falls into four broad challenges, i.e., administrative challenges, academic challenges, operational difficulties, and environmental problems. Each of these challenges requires a very intensive effort to tackle. KAU administration has significantly contributed to providing solutions to all the challenges faced by the program, as a result of the new establishment of the program.

5.1.1. Administrative

The financial procedures have represented one of the biggest challenges regarding allocating budgets, and the spending in the newly established hydrographic surveying program is very generous. However, the financial routines are time consuming and take the time to complete, which delays the program in the beginning. King Abdulaziz University, with support from the Ministry of Higher Education, has contributed to solving this problem by allocating a budget for the new project, that is independent of the budget of the Faculty of Marine Science, the host of the new faculty. In fact, the budget for the new project has in many ways exceeded the budget of the hosting faculty. This has contributed to streamlining the process on spending on facilities as equipment for the new faculty.

Another challenge was characterized by the ambiguity in decision-making, in the form of having a new department with four subspecialties administratively under Marine Science. The department is following the standard decision-making process in the university. This decision-making process for the new department completed under the approval of the Faculty of Marine Science. King Abdulaziz University has supported independent decision-making process for the newly established department, specifically in financial administration and hiring processes. This decision-making model has provided the flexibility to the organization structure of subdepartments with professional identity located administratively under the Faculty of Marine Science to work more dynamically. However, issues related to students records, registration, and all student affairs were handled under the Faulty of Marine Science, and it was sometimes delaying factor in following the administrative hierarchy. The dual identity of the Department of Maritime Studies, in a sense that the budget of the department is independent, but the organizational hierarchy somewhat falls under the Faculty of Maritime Studies, has created some confusion in the administration in the university. Many contradictions in policies reflected the dual identity related to the Department of Maritime Studies as an individual department, which hold the seeds for new faculty. The higher administration of King Abdulaziz University has exercised every effort to facilitate the transition of the new Faculty of Maritime Studies. In 2011, the Faculty of Maritime Studies was an independent department with a royal decree that announced its existence for the first time.

Another administrative challenge was related to hiring and retention of faculty members in hydrographic surveying. In general, all geomatics specializations are considered rare specialties. This global shortage situation, also, limited spending controlled by outdated policies and procedures as well as the lack of local Saudis faculty members, and staff has made KAU hydrographic surveying program relatively unstable, in the first years. King Abdulaziz University has facilitated this process by activating extra incentive within the pay scheme approved by the Ministry of Higher Education, specifically for attracting faculty members in unique specializations. Also, KAU has provided additional incentive and local grants for research projects that support faculty research activities in small or medium projects.

On the program implementation side, King Abdulaziz University strived to provide world class geomatics education; however, there was some skepticism from students to put their future in the line while applying for a new program. Issues related to field training were among the most significant challenge to this program and for few years in the beginning. Students had to wait for few months for placement in field training. The university administration has realized the importance of providing the students with a perfect academic experience and attempted to support that in two ways. The first way was through allocating generous budgets for facilities and equipment, including the purchase of advanced hydrographic surveying vessel with all equipment needed for training and with onboard lab. The second way was by providing the students with short internship training through several local and international memorandum of understanding (MOUs) with well-reputed professional organizations, including the Saudi Aramco and some of its contractors, who have accepted to train some students in the field over the summer term. Additionally, agreements were signed with international training centers in the Netherlands, France, and Canada. Different batches of students sponsored by KAU to complete their summer training term abroad to get exposed to international operation environments since 2009 groups of students completed their training in the Netherlands, France, and Canada for training. Despite the arrival of the department training vessel in 2010, the administration of King Abdulaziz University has decided to keep the international training program running to maximize the students' international training experience and to provide international collaboration with counterpart institutions in Europe and North America. The training for 2017 is planned to be in the United States.

5.1.2. Operational

On the operational side, some challenges have faced the newly established hydrographic surveying program. Summary of these difficulties fall in equipment and infrastructure; as the program is new, its facilities and infrastructures are still being developed. Most of lab hardware and software capabilities were developing. As a newly established department, the department was faced with the issue of filling all vacant posts from technical staff to help faculty and administration with different operational issues including communications and archiving. King Abdulaziz University has assisted in this regard by providing some positions for hiring new technical staff, and all posts are now filled, with many specialized hydrographers to support with onboard the vessel training and in the laboratories. Also, additional support staff was hired to take responsibilities for educational affairs and communications.

Historically, students were faced with the challenge of accommodating schedules dispersed over two campuses, 40 km apart. It was a major operational challenge having lectures before noon on the main campus and scheduled labs in the afternoon in the Al Morgan campus. However, King Abdulaziz University and the faculty administration have arranged to bring all classes on the same campus in Al Morgan, saving significant time and effort for students to commute between the two campuses.

5.1.3. Evaluation of learning outcomes

The program is providing new technology-based education in the region. Students were trained to gain competence in sophisticated equipment and hardware. The quality of the program was discussed based on indicators other than the self-reported data from the students. These indicators of the program's success and positive are the employability of the graduates and securing international recognition as a successful professionals, and IHO Cat A certified professionals. These are key indicators of developing high-quality local professionals that are demand for graduates. Reports from international training suggested that our graduates' skills and knowledge are adequate to allow them to gain advanced training skills abroad, specifically in Canada. The students have utilized the program capabilities and facilities efficiently and were able to get jobs easily. Seven of our graduates were employed by the department, and five of them have completed masters in the United States, Canada, the Netherlands, the United Kingdom, and France. Currently, three of our program graduates are pursuing Ph.D. studies in Australia, the United Kingdom, and the United States to take over as faculty members in future.

5.2. Development of sociocultural indicators

The socio-cultural contributions of the program are evident to the geomatics community in Saudi Arabia and the region. The international impact of the program was evident in the input of the program faculty in research projects locally and internationally. **Table 3** is showing the employment rate of the graduates of the program between 2007 and 2016. **Table 4** below is showing the contributions of the program faculty in the last 5 years. The trend is growing as new research projects development is increasing. The department is in active collaboration

Year	Number of graduates	Employment rate	Employer
2007	3	100%	King Abdulaziz University, Saudi Geological Survey, private sector
2008	7	100%	
2009	3	100%	Saudi Aramco, private sector, King Abdulaziz University
2010	5	100%	King Abdulaziz University General Commission for Survey, Armed Forces, Saudi Aramco
2011	7	100%	Private sector
2013	3	100%	Private sector
2014	5	100%	Private sector
2015	8	100%	General Commission for Survey
2016	10	100%	Private sector

Table 3. Students employment history for graduates between 2007 and 2016.

Year	International peer-reviewed journal publications	International conference presentations
2012	4	0
2013	5	2
2014	6	1
2015	6	3
2016	7	1

Table 4. International contributions of the program in the last 5 years.

with York University and Concordia University in Canada in ongoing research projects, and there is an agreement with the Center for Interdisciplinary Development in Ocean Mapping (CIDCO) in Canada for the training of students.

The program has contributed to the economic development of the region by providing highly skilled professionals that are serving in many key employers in the public and private sectors. The program started to familiarize local decision-makers as well as the public with the role and capabilities of geomatics in general and hydrography specifically. The program was visited by a delegation of the Hydrographic Society of the United Kingdom, UAE branch to extend an invitation to the program faculty, and students to contribute to the outreach activities in the region. The program outreach committee has invited some high schools to schedule visits of their final year students to the department and vessel, where students visited the program and learned about the capabilities the program and the requirements for admission. Also, the department has participated in many local conferences and introduced the program at different capacities for professionals as well as for nontechnical extended community members.

Many high school students showed interest in the program, and some had participated in mini-projects involving hydrographic aspects, where they stayed in contact with the department while working on projects. A delegation of the High School Teachers Conference in the GCC region has visited the department and learned about hydrographic surveying equipment and vessel. The local media, national TV, and other private channels have highlighted several activities by the program, bringing knowledge about the importance of hydrographic surveying.

The program is contributing to the development of a unique community of hydrographers in the Kingdom of Saudi Arabia and the region. It is getting support from KAU administration to extend its role in community service through different educational and awareness activities, as well as through developed short-training courses taught at some institutions in the City of Jeddah.

6. Conclusions

This chapter discussed a collaborative approach in determining the challenges faced by a newly established hydrographic surveying department at King Abdulaziz University, Saudi Arabia. It has presented the opportunities for developing such a program through transfer and modification of the established approach, using KAU program as a model. The newly created program addressed the apparent need for such specialized and targeted higher education that provides international professional certification, especially by looking at the environmental issues in the region, natural resources and maritime transport in the area, and the need for safe navigation. The study has demonstrated the unique inception of a program that meets the international standards, while providing high-quality local education for students in the region has contributed to the socioeconomic well-being of the region and can be seen as a model for similar future programs worldwide.

The establishment challenges faced by the program can be better addressed by proposing a new scheme of collaboration and by exploring new short-term models of recruitment that attract visiting faculty members from well-reputed universities for short-term visiting professorship trips, or by signing partnership agreements with the different universities mainly for faculty exchange and training of students. The support from the international community represented by the IHO and host training institutions was among the key attracting points from various stakeholders in the Saudi Arabia and the GCC region broadly. The program has provided a new, world-class hydrographic surveying education through an effective partnership with the IHO, IHO-IBSC, and the leading training institutions in Canada, the Netherlands, and France. The satisfaction of the program stakeholders and the success of the program to provide employment opportunities to the graduates were a direct measure of the success of the program in meeting the stakeholders' expectations. Over the past 10 years, the graduates' employment rate within 6 months of graduation was 100%. Several meetings with the stakeholders reflected the high satisfaction of the program capabilities and the graduates' preparation. The Saudi Aramco, the leading oil company in the world, is a strategic partner of the program. The management of the surveying division of Saudi Aramco has expressed interest to hire all graduates in the upcoming through the partnership agreement signed with the Faculty of Maritime Studies. Stakeholders' enthusiasm and KAU higher administration strong support were key factors in the firm inception of the program, backed by generous budgets, resources, and passion for accomplishment. The program has also developed new opportunities for the stakeholders to expand their hydrographic surveying capabilities. GCS has hired all graduates of the 2014 class and started to expand the hydrographic surveying capabilities of the authority.

Author details

Rifaat Abdalla[1]* and Salim Al-Harbi[2]

*Address all correspondence to: rabdalla@squ.edu.om

1 Department of Earth Sciences, College of Science, Sultan Qaboos University, Oman

2 Department of Hydrographic Surveying, Faculty of Maritime Studies, King Abdulaziz University, Jeddah, Saudi Arabia

References

[1] Natural Resources Canada. Geomatics Sector. 2014. Available from: http://www.nrcan.gc.ca/earth-sciences/geomatics/10776 [Accessed: July 1, 2014]

[2] Bekesi E. GIS education in New Zealand tertiary institutions-information systems perspective. In: Paper Presented at the 8th Colloquium of the Spatial Information Research Center, Otago, Switzerland. 1996

[3] Sifakis J. A vision for computer science—The system perspective. Central European Journal of Computer Science. 2011;1(1):108-116

[4] Lilian P-C, Tammy K. The development of geomatics education in Hong Kong, Vol. 10: 4. International Research in Geographical and Environmental Education. 2001;10(1):88-91

[5] Konecny G. Recent global challenge in geomatics education. International Archives of the Photogrammetry, Remote Sensing, and Information Sciences. 2002;XXXV(6):6

[6] Roche S, Caron C. Geomatics solutions not universal: From a western to a central/eastern European perspective. GIM International. 2001;15(6):33-35

[7] Robert W, John H. Spatial Science Education Directions for QUT. In: Les F, Editor. Proceedings Combined 5th Trans Tasman Survey Conference and the 2nd Queensland Spatial Industry Conference, Cairns, Queensland, Australia. 2006. pp. 00037-00037. Accessed from: http://eprints.qut.edu.au

[8] Case JM, Light G. The case. Emerging methodologies in engineering education research. Journal of Engineering Education. 2011;**100**(1):186-210

[9] Cousin G. Researching Learning in Higher Education: An Introduction to Contemporary Methods and Approaches. New York: Routledge; 2009

[10] Johri A, Olds BM. Situated engineering learning: Bridging engineering education research and the learning sciences. Journal of Engineering Education. 2011;**100**(1): 151-185. DOI: 10.1002/j.2168-9830.2011.tb00007.x

[11] Agrafiotis D, Koumoutsos N. Problems of general and technical education. European Journal of Engineering Education. 1981;**6**(1-2):147-158. DOI: 10.1080/03043798108547805

[12] Duncan EE. Geomatics education in Ghana. In: Paper Presented at the FIG Working Week 2004, Athens, Greece, May 22-27, 2004

[13] Young FR. Geomatics education—A modern day. Australian Surveyor. 1992;**37**(4):289-294

[14] Fajemirokun FA, Nwilo PC, Badejo OT. Geomatics education in Nigeria. In: Paper Presented at the FIG XXII international congress, Washington, D.C. USA, April 19-26, 2002. 2002

[15] Aina YA. The need, and development of geomatics education in Saudi Arabia. In: The Second National GIS Symposium in Saudi Arabia, Dharan, April 2007. 2007. www.sau-digis.org

[16] Akinyemi FO. Technology transfer: Assessing the impact of desktop cartography course on mapping professionals in Nigeria (1998-2001). In: Paper Presented at The International Archives of the Photogrammetry, Remote Sensing, and Spatial Information Sciences. 2002

[17] Abdalla R. Development of hydrographic surveying database for the Red Sea. In: Paper Presented at The International Hydrographic Organization (IHO) and the Regional Organization for the Conservation of the Red Sea and The Gulf of Aden (PERSGA) Regional Capacity Building Workshop, Jeddah, November 10-23, 2007. 2007

[18] Höhle J. Project-based learning in geomatics. In: International Archives of the Photogrammetry, Remote Sensing and Spatial Information Science, Tokyo, Japan, 2006. Vol. XXXVI. Berlin, Germany: ISPRS; 2005. pp. 1-8

[19] Burkholder EF. Geomatics curriculum design issues. Surveying and Land Information Science. 2005;**65**(3):151-157

[20] Slater F, Graves N, Lambert D. Editorial. International Research in Geographical and Environmental Education. 2016;**25**(3):189-194. DOI: 10.1080/10382046.2016.1155321

[21] Heinz R. The situation of geomatics education in Africa–An endangered profession. International Federation of Surveyors (article of the month). 2003:1-13

[22] Aina YA. Geomatics education in Saudi Arabia: Status, challenges, and prospects. International Research in Geographical and Environmental Education. 2009;**18**(2):111-119. DOI: 10.1080/10382040902861197

[23] Melezinek A. Technology and its education. Some approach and experiences from Austria. European Journal of Engineering Education. 1982;7(1):61-66. DOI: 10.1080/03043798208903637

[24] Bingham GA, Southee DJ, Page T. Meeting the expectation of industry: An integrated approach to the teaching of mechanics and electronics to design students. European Journal of Engineering Education. 2015;40(4):410-431. DOI: 10.1080/03043797.2014.1001813

[25] Nile G, Wang J, Gau J-T. Challenges in teaching modern manufacturing technologies. European Journal of Engineering Education. 2015;40(4):432-449. DOI: 10.1080/03043797.2014.1001814

[26] Venugopal K, Senthil R, Yogendran S. Geomatics education in India—A view point. In: Paper Presented at the 22 Asian Conference on Remote Sensing, Singapore, November 5-9, 2001. 2001

[27] Vikki-Hatakka T, Tuunila R, Nurkka N. Development of chemical engineering course methods using action research: A case study. European Journal of Engineering Education. 2013;38(5):469-484. DOI: 10.1080/03043797.2013.811471

[28] International Hydrographic Organization. Standards of competence for hydrographic surveyors. In: IHB, editor. Guidance and Syllabus for Educational and Training Programmes. 11th ed. Monaco: IHO Publication S5; 2011. p. 65

[29] Kufoniyi O, Huurneman G, Horn J. Human and institutional capacity building in geoinformatics through educational networking. In: Paper Presented at the From Pharaohs to Geoinformatics, FIG Working Week, Eight the Global Spatial Data Infrastructure Conference, Cairo, Egypt, April 16-21, 2005. 2005

[30] KAU. Hydrographic Surveying Curriculum. Jeddah, KSA: KAU Press; 2007. p. 192

[31] Rashid M, Tasadduq I. Holistic development of computer engineering curricula using Y-chart methodology. IEEE Transactions on Education. 2014;57(3):193-200

[32] Saudi Council of Engineers (SCE). Accreditation Policies. 2006. Available from: http://www.saudieng.sa/English/pages/default.aspx

Mathematical Analysis of Some Typical Problems in Geodesy by Means of Computer Algebra

Hou-pu Li and Shao-feng Bian

Additional information is available at the end of the chapter

http://dx.doi.org/10.5772/intechopen.81586

Abstract

There are many complicated and fussy mathematical analysis processes in geodesy, such as the power series expansions of the ellipsoid's eccentricity, high order derivation of complex and implicit functions, operation of trigonometric function, expansions of special functions and integral transformation. Taking some typical mathematical analysis processes in geodesy as research objects, the computer algebra analysis are systematically carried out to bread, deep and detailed extent with the help of computer algebra analysis method and the powerful ability of mathematical analysis of computer algebra system. The forward and inverse expansions of the meridian arc in geometric geodesy, the nonsingular expressions of singular integration in physical geodesy and the series expansions of direct transformations between three anomalies in satellite geodesy are established, which have more concise form, stricter theory basis and higher accuracy compared to traditional ones. The breakthrough and innovation of some mathematical analysis problems in the special field of geodesy are realized, which will further enrich and perfect the theoretical system of geodesy.

Keywords: geodesy, computer algebra, mathematical analysis, meridian arc, singular integration, mean anomaly

1. Introduction

Geodesy is the science of accurately measuring and understanding three fundamental properties of the Earth: its geometric shape, its orientation in space, and its gravity field, as well as the changes of these properties with time. There are many fussy symbolic problems to be dealt with in geodesy, such as the power series expansions of the ellipsoid's eccentricity, high order derivation of complex and implicit functions, expansions of special functions and integral

IntechOpen

transformation. Many geodesists have made great efforts to solve these problems, see [1–8]. Due to historical condition limitation, they mainly disposed of these problems by hand, which were not perfectly solved yet. Traditional algorithms derived by hand mainly have the following problems: (1) The expressions are complex and lengthy, which makes the computation process very complicated and time-consuming. (2) Some approximate disposal is adopted, which influences the computation accuracy. (3) Some formulae are numerical and only apply to a specific reference ellipsoid, which are not convenient to be generalized.

In computational mathematics, computer algebra, also called symbolic computation, is a scientific area that refers to the study and development of algorithms and software for manipulating mathematical expressions and other mathematical objects. Software applications that perform symbolic calculations are called computer algebra systems, which are more popular today. Computer algebra systems, like Mathematica, Maple and Mathcad, are powerful tools to solve some mathematical derivation in geodesy, see [9–11]. By means of computer algebra analysis method and computer algebra system Mathematica, we have already solved many complicated mathematical problems in special fields of geodesy in the past few years; see [12–15].

The main contents and research results presented in this chapter are organized as follows: In Section 2, we discuss the forward and inverse expansions of the meridian arc often used in geometric geodesy. In Section 3, the nonsingular expressions of singular integration in physical geodesy are derived. In Section 4, we discuss series expansions of direct transformations between three anomalies in satellite geodesy. Finally in Section 5, we make a brief summary.

2. The forward and inverse expansions of the meridian arc in geometric geodesy

The forward and inverse problem of the meridian arc is one of the fundamental problems in geometric geodesy, which seems to be a solved one. Briefly reviewing the existing methods, however, one will find that the inverse problem was not perfectly and ideally solved yet. This situation is due to the complexity of the problem itself and the lack of advanced computer algebra systems. Yang had given the direct expansions of the inverse transformation by means of the Lagrange series method, but their coefficients are expressed as polynomials of coefficients of the forward expansions, which are not convenient for practical use, see [6, 7]. Adams expressed the coefficients of inverse expansions as a power series of the eccentricity e by hand, but expanded them up to eighth-order terms of e at most, see [1]. Due to these reasons, the forward and inverse expansions of the meridian arc are discussed by means of Mathematica in the following sections. Their coefficients are uniformly expressed as a power series of the eccentricity and extended up to tenth-order terms of e.

2.1. The forward expansion of the meridian arc

The meridian arc from the equator where the latitude is from $B = 0$ to B is

$$X = a(1 - e^2) \int_0^B (1 - e^2 \sin^2 B)^{-3/2} dB \qquad (1)$$

where X is the meridian arc; B is the geodetic latitude; a is the semi-major axis of the reference ellipsoid; e is the first eccentricity of the reference ellipsoid.

Expanding the integrand in Eq. (1) and integrating it item by item using Mathematica as follows:

$$S = \text{Series}\left[\left(1 - e^2 * \text{Sin}[B]^2\right)^{-3/2}, \{e, 0, 10\}\right]$$

$$1 + \frac{3}{2}\,\text{Sin}[B]^2\,e^2 + \frac{15}{8}\,\text{Sin}[B]^4\,e^4 + \frac{35}{16}\,\text{Sin}[B]^6\,e^6 + \frac{315}{128}\,\text{Sin}[B]^8\,e^8 + \frac{693}{256}\,\text{Sin}[B]^{10}\,e^{10} + O[e]^{11}$$

$$X0 = \text{Integrate}[S, B]$$

$$B + \frac{3}{4}\,(B - \text{Cos}[B]\,\text{Sin}[B])\,e^2 + \frac{15}{256}\,(12\,B - 8\,\text{Sin}[2\,B] + \text{Sin}[4\,B])\,e^4 -$$
$$\frac{35\,(-60\,B + 45\,\text{Sin}[2\,B] - 9\,\text{Sin}[4\,B] + \text{Sin}[6\,B])\,e^6}{3072} + \frac{1}{131\,072}$$
$$105\,(840\,B - 672\,\text{Sin}[2\,B] + 168\,\text{Sin}[4\,B] - 32\,\text{Sin}[6\,B] + 3\,\text{Sin}[8\,B])\,e^8 + \frac{1}{2\,621\,440}$$
$$693\,(2520\,B - 2100\,\text{Sin}[2\,B] + 600\,\text{Sin}[4\,B] - 150\,\text{Sin}[6\,B] + 25\,\text{Sin}[8\,B] - 2\,\text{Sin}[10\,B])\,e^{10} + O[e]^{11}$$

$$X0 = \text{TrigReduce}[\text{Normal}[X0]]$$

$$\frac{1}{7\,864\,320}$$
$$(7\,864\,320\,B + 5\,898\,240\,B\,e^2 + 5\,529\,600\,B\,e^4 + 5\,376\,000\,B\,e^6 + 5\,292\,000\,B\,e^8 + 5\,239\,080\,B\,e^{10} -$$
$$2\,949\,120\,e^2\,\text{Sin}[2\,B] - 3\,686\,400\,e^4\,\text{Sin}[2\,B] - 4\,032\,000\,e^6\,\text{Sin}[2\,B] - 4\,233\,600\,e^8\,\text{Sin}[2\,B] -$$
$$4\,365\,900\,e^{10}\,\text{Sin}[2\,B] + 460\,800\,e^4\,\text{Sin}[4\,B] + 806\,400\,e^6\,\text{Sin}[4\,B] + 1\,058\,400\,e^8\,\text{Sin}[4\,B] +$$
$$1\,247\,400\,e^{10}\,\text{Sin}[4\,B] - 89\,600\,e^6\,\text{Sin}[6\,B] - 201\,600\,e^8\,\text{Sin}[6\,B] -$$
$$311\,850\,e^{10}\,\text{Sin}[6\,B] + 18\,900\,e^8\,\text{Sin}[8\,B] + 51\,975\,e^{10}\,\text{Sin}[8\,B] - 4158\,e^{10}\,\text{Sin}[10\,B])$$

Then one arrives at

$$X = a\left(1 - e^2\right)\left(K_0 B + K_2 \sin 2B + K_4 \sin 4B + K_6 \sin 6B + K_8 \sin 8B + K_{10} \sin 10B\right) \tag{2}$$

where

$$\begin{cases} K_0 = 1 + \frac{3}{4}e^2 + \frac{45}{64}e^4 + \frac{175}{256}e^6 + \frac{11025}{16384}e^8 + \frac{43659}{65536}e^{10} \\[2mm] K_2 = -\frac{3}{8}e^2 - \frac{15}{32}e^4 - \frac{525}{1024}e^6 - \frac{2205}{4096}e^8 - \frac{72765}{131072}e^{10} \\[2mm] K_4 = \frac{15}{256}e^4 + \frac{105}{1024}e^6 + \frac{2205}{16384}e^8 + \frac{10395}{65536}e^{10} \\[2mm] K_6 = -\frac{35}{3072}e^6 - \frac{105}{4096}e^8 - \frac{10395}{262144}e^{10} \\[2mm] K_8 = \frac{315}{131072}e^8 + \frac{3465}{524288}e^{10} \\[2mm] K_{10} = -\frac{693}{1310720}e^{10} \end{cases} \tag{3}$$

Eqs. (2) and (3) can also be derived using the binomial theorem by hand which consumes much more time, see [6–9]. The denominator 693 of the last coefficient K_{10} is mistaken as 639 in Ref. [9].

2.2. The inverse expansion of the meridian arc using the Hermite interpolation method

Differentiation to the both sides of Eq. (1) yields:

$$\frac{dX}{dB} = \frac{a(1 - e^2)}{(1 - e^2 \sin^2 B)^{3/2}}$$ (4)

Define ψ as

$$\psi = \frac{X}{a(1 - e^2)K_0}$$ (5)

Substituting Eq. (5) into Eq. (4) yields:

$$\frac{dB}{K_0 d\psi} = (1 - e^2 \sin^2 B)^{3/2}$$ (6)

Suppose that the inverse solution of Eq. (6) has the following form:

$$B = \psi + a_2 \sin 2\psi + a_4 \sin 4\psi + a_6 \sin 6\psi + a_8 \sin 8\psi + a_{10} \sin 10\psi$$ (7)

Eq. (7) has five coefficients to be determined. Once these coefficients are known, the inverse problem can be solved.

We consider that the values of differentiation Eq. (6) at the beginning and end points can be treated as interpolation constraints. It is generally known that

$$B'(0) = K_0$$ (8)

$$B'\left(\frac{\pi}{2}\right) = K_0(1 - e^2)^{3/2}$$ (9)

The further derivation of Eq. (6) with respect to ψ yields $B''(\psi)$. Unfortunately, $B''(\psi)$ is equal to zero at $\psi = 0$, $\psi = \frac{\pi}{2}$. Hence, one differentiates Eq. (6) twice and it yields $B'''(\psi)$. Omitting the derivative procedure by means of Mathematica, one arrives at

$$B'''(0) = -3 e^2 - \frac{27}{4} e^4 - \frac{729}{64} e^6 - \frac{4329}{256} e^8 - \frac{381645}{16384} e^{10}$$ (10)

$$B'''\left(\frac{\pi}{2}\right) = 3 e^2 - \frac{15}{4} e^4 + \frac{57}{64} e^6 + \frac{3}{256} e^8 - \frac{51}{16384} e^{10}$$ (11)

The further derivation of $B'''(\psi)$ with respect to ψ yields $B^{(4)}(\psi)$, but $B^{(4)}(\psi)$ is equal to zero at $\psi = 0$, $\psi = \frac{\pi}{2}$. Hence, one differentiates $B'''(\psi)$ twice and it yields $B^{(5)}(\psi)$. Then one arrives at

$$B^{(5)}(0) = 12e^2 + 90e^4 + \frac{4455}{16}e^6 + \frac{20145}{32}e^8 + \frac{4924935}{4096}e^{10} \tag{12}$$

Making use of the five interpolation constraints in Eqs. (8–12) and differentiating Eq. (7) correspondingly, one arrives at a set of linear equations for the unknown coefficients

$$
\begin{pmatrix}
2 & 4 & 6 & 8 & 10 \\
-2 & 4 & -6 & 8 & -10 \\
-8 & -64 & -216 & -512 & -1000 \\
8 & -64 & -216 & -512 & 1000 \\
32 & 1024 & 7776 & 32768 & 100000
\end{pmatrix}
\begin{pmatrix}
a_2 \\ a_4 \\ a_6 \\ a_8 \\ a_{10}
\end{pmatrix}
=
\begin{pmatrix}
B'(0) - 1 \\
B'\left(\frac{\pi}{2}\right) - 1 \\
B''(0) \\
B''\left(\frac{\pi}{2}\right) \\
B^{(5)}(0)
\end{pmatrix}
\tag{13}
$$

The solution to Eq. (13) is

$$
\begin{pmatrix}
a_2 \\ a_4 \\ a_6 \\ a_8 \\ a_{10}
\end{pmatrix}
=
\begin{pmatrix}
2 & 4 & 6 & 8 & 10 \\
-2 & 4 & -6 & 8 & -10 \\
-8 & -64 & -216 & -512 & -1000 \\
8 & -64 & -216 & -512 & 1000 \\
32 & 1024 & 7776 & 32768 & 100000
\end{pmatrix}^{-1}
\begin{pmatrix}
B'(0) - 1 \\
B'\left(\frac{\pi}{2}\right) - 1 \\
B''(0) \\
B''\left(\frac{\pi}{2}\right) \\
B^{(5)}(0)
\end{pmatrix}
\tag{14}
$$

Omitting the main operations by means of Mathematica, one arrives at

$$
\begin{cases}
a_2 = \dfrac{3}{8}e^2 + \dfrac{3}{16}e^4 + \dfrac{213}{2048}e^6 + \dfrac{255}{4096}e^8 + \dfrac{20861}{524288}e^{10} \\[2mm]
a_4 = \dfrac{21}{256}e^4 + \dfrac{21}{256}e^6 + \dfrac{533}{8192}e^8 + \dfrac{197}{4096}e^{10} \\[2mm]
a_6 = \dfrac{151}{6144}e^6 + \dfrac{151}{4096}e^8 + \dfrac{5019}{131072}e^{10} \\[2mm]
a_8 = \dfrac{1097}{131072}e^8 + \dfrac{1097}{65536}e^{10} \\[2mm]
a_{10} = \dfrac{8011}{2621440}e^{10}
\end{cases}
\tag{15}
$$

2.3. The inverse expansion of the meridian arc using Lagrange's theorem method

Suppose that

$$y = x + f(x) \tag{16}$$

with $|f(x)| \ll |x|$ and $y \approx x$. Lagrange's theorem states that in a suitable domain the solution of Eq. (16) is

$$x = y + \sum_{n=1}^{\infty} \frac{(-1)^n}{n!} \frac{d^{n-1}}{dy^{n-1}} [f(y)]^n \tag{17}$$

Suppose $f(x)$ is defined by the following finite trigonometric series:

$$f(x) = \alpha \sin 2x + \beta \sin 4x + \gamma \sin 6x + \delta \sin 8x + \varepsilon \sin 10x \tag{18}$$

where the coefficients $\alpha = O(e^2)$, $\beta = O(e^4)$, $\gamma = O(e^6)$, $\delta = O(e^8)$, $\varepsilon = O(e^{10})$ are small enough for the condition $|f(x)| \ll |x|$. In deriving the inversion we shall truncate the infinite Lagrange expansion at eighth-order terms of e and drop higher powers. Inserting Eq. (18) into Eq. (17), one arrives at

$$x = y - f(y) + \frac{1}{2!} \frac{d}{dy} [f(y)]^2 - \frac{1}{3!} \frac{d^2}{dy^2} [f(y)]^3 + \frac{1}{4!} \frac{d^3}{dy^3} [f(y)]^4 - \frac{1}{5!} \frac{d^4}{dy^4} [f(y)]^5 \tag{19}$$

One should make use of several trigonometric identities to calculate the derivatives, substituting them into Eq. (19) and grouping terms according to the trigonometric functions. It is a difficult and time-consuming work to do by hand, but could be easily realized by means of Mathematica. Omitting the main procedure, one arrives at

$$x = y + d_2 \sin 2y + d_4 \sin 4y + d_6 \sin 6y + d_8 \sin 8y + d_{10} \sin 10y \tag{20}$$

where

$$\begin{cases} d_2 = -\alpha - \alpha\beta - \beta\gamma + \frac{1}{2}\alpha^3 + \alpha\beta^2 - \frac{1}{2}\alpha^2\gamma + \frac{1}{3}\alpha^3\beta - \frac{1}{12}\alpha^5 \\[2mm] d_4 = -\beta + \alpha^2 - 2\alpha\gamma + 4\alpha^2\beta - \frac{4}{3}\alpha^4 \\[2mm] d_6 = -\gamma + 3\alpha\beta - 3\alpha\delta - \frac{3}{2}\alpha^3 + \frac{9}{2}\alpha\beta^2 + 9\alpha^2\gamma - \frac{27}{2}\alpha^3\beta + \frac{27}{8}\alpha^5 \\[2mm] d_8 = -\delta + 2\beta^2 + 4\alpha\gamma - 8\alpha^2\beta + \frac{8}{3}\alpha^4 \\[2mm] d_{10} = -\varepsilon + 5\alpha\delta + 5\beta\gamma - \frac{25}{2}\alpha\beta^2 - \frac{25}{2}\alpha^2\gamma + \frac{125}{6}\alpha^3\beta - \frac{125}{24}\alpha^5 \end{cases} \tag{21}$$

Substituting x for B and y for ψ in Eq. (16), the coefficients α, β, γ, δ, ε are consistent with α_2, α_4, α_6, α_8 in Eq. (3). According to Eq. (20) and denoting a_2, a_4, a_6, a_8, a_{10} as the new coefficients, the inverse expansion of the meridian arc can be written as

$$B = \psi + a_2 \sin 2\psi + a_4 \sin 4\psi + a_6 \sin 6\psi + a_8 \sin 8\psi + a_{10} \sin 10\psi \tag{22}$$

where

$$
\begin{cases}
a_2 = -\alpha_2 - \alpha_2\alpha_4 - \alpha_4\alpha_6 + \dfrac{1}{2}\alpha_2^3 + \alpha_2\alpha_4^2 - \dfrac{1}{2}\alpha_2^2\alpha_6 + \dfrac{1}{3}\alpha_2^3\alpha_4 - \dfrac{1}{12}\alpha_2^5 \\[2mm]
a_4 = -\alpha_4 + \alpha_2^2 - 2\alpha_2\alpha_6 + 4\alpha_2^2\alpha_4 - \dfrac{4}{3}\alpha_2^4 \\[2mm]
a_6 = -\alpha_6 + 3\alpha_2\alpha_4 - 3\alpha_2\alpha_8 - \dfrac{3}{2}\alpha_2^3 + \dfrac{9}{2}\alpha_2\alpha_4^2 + 9\alpha_2^2\alpha_6 - \dfrac{27}{2}\alpha_2^3\alpha_4 + \dfrac{27}{8}\alpha_2^5 \\[2mm]
a_8 = -\alpha_8 + 2\alpha_4^2 + 4\alpha_2\alpha_6 - 8\alpha_2^2\alpha_4 + \dfrac{8}{3}\alpha_2^4 \\[2mm]
a_{10} = -\alpha_{10} + 5\alpha_2\alpha_8 + 5\alpha_4\alpha_6 - \dfrac{25}{2}\alpha_2\alpha_4^2 - \dfrac{25}{2}\alpha_2^2\alpha_6 + \dfrac{125}{6}\alpha_2^3\alpha_4 - \dfrac{125}{24}\alpha_2^5
\end{cases}
\tag{23}
$$

The coefficients in Eq. (23) are also easily expressed in a power series of the eccentricity by means of Mathematica. Omitting the main procedure, one arrives at

$$
\begin{cases}
a_2 = \dfrac{3}{8}e^2 + \dfrac{3}{16}e^4 + \dfrac{213}{2048}e^6 + \dfrac{255}{4096}e^8 + \dfrac{20861}{524288}e^{10} \\[2mm]
a_4 = \dfrac{21}{256}e^4 + \dfrac{21}{256}e^6 + \dfrac{533}{8192}e^8 + \dfrac{197}{4096}e^{10} \\[2mm]
a_6 = \dfrac{151}{6144}e^6 + \dfrac{151}{4096}e^8 + \dfrac{5019}{131072}e^{10} \\[2mm]
a_8 = \dfrac{1097}{131072}e^8 + \dfrac{1097}{65536}e^{10} \\[2mm]
a_{10} = \dfrac{8011}{2621440}e^{10}
\end{cases}
\tag{24}
$$

The results in Eq. (24) are consistent with the coefficients in Eq. (15), which substantiates the correctness of the derived formula.

2.4. The accuracy of the inverse expansion of the meridian arc

In order to validate the exactness of the inverse expansions of meridian arc derived by the author, one has examined its accuracy choosing the CGCS2000 reference ellipsoid with parameters $a = 6378137$, $e = 0.08181919104281579$. The accuracy of the inverse expansions derived by Yang (see [6, 7]) is also examined for comparison.

One makes use of Eq. (1) and Eq. (5) to obtain the theoretical value ψ_0 at given geodetic latitude B_0. Then one makes use of the inverse expansions derived by Yang (see [6, 7]) to obtain the computation value B_1. Substituting ψ_0 into Eq. (22), one arrives at the computation value B_1'. The differences $\Delta B_i = B_i - B_0$, $\Delta B_i' = B_i' - B_0 (i = 1, 2)$ indicate the accuracies of the inverse expansions derived by Yang (see [6, 7]) and the author respectively. These errors are listed in **Table 1**.

$B_0/(°)$	20	40	60	80
$\Delta B_1/('')$	-3.2×10^{-7}	-1.6×10^{-6}	-1.8×10^{-6}	-1.4×10^{-6}
$\Delta B_1'/('')$	2.7×10^{-9}	8.6×10^{-9}	1.3×10^{-8}	1.7×10^{-8}

Table 1. Errors of the inverse expansions of meridian arc.

From **Table 1**, one could find that the accuracy of the inverse expansion of meridian arc derived by Yang (see [6, 7]) is higher than $10^{-5''}$, and the accuracy of the inverse expansion Eq. (22) derived by the author is higher than $10^{-7''}$. The accuracy is improved by 2 orders of magnitude by means of computer algebra.

3. The singular integration in physical geodesy

Singular integrals associated with the reciprocal distance usually exist in the computations of physical geodesy and geophysics. For example, the integral expressions of height anomaly, deflections of the vertical and vertical gradient of gravity anomaly can be written in planar approximation as

$$\zeta = \frac{1}{2\pi\gamma} \iint \frac{\Delta g}{r} dxdy \tag{25}$$

$$\xi = -\frac{1}{2\pi\gamma} \iint \frac{x\Delta g}{r^3} dxdy \tag{26}$$

$$\eta = -\frac{1}{2\pi\gamma} \iint \frac{y\Delta g}{r^3} dxdy \tag{27}$$

$$L = \frac{1}{2\pi} \iint \frac{\Delta g - \Delta g_0}{r^3} dxdy \tag{28}$$

where $r = \sqrt{x^2 + y^2}$, Δg_0 is the gravity anomaly at the computation point. When $r \to 0$, the above integrals become singular and need special treatment at the computation point. The past treatments are with respect to template computations, which regards the innermost area as a circle, see [3, 16]. But the gravity anomalies are given in the rectangular grids(such as $2' \times 2'$), if the approximate disposal is used, some significant error may be introduced. Sunkel and Wang expressed the gravity anomalies block by block as an interpolation polynomial and derived the analytic values of the integrals, see [17, 18]. However, the integrals of the rational functions are very complicated, especially when related interpolation polynomials contain many terms. One only can give the analytic values of the corresponding linear approximation. In this chapter, we use the nonsingular integration transformations proposed by Bian (see [19]) to compute the above integrals precisely.

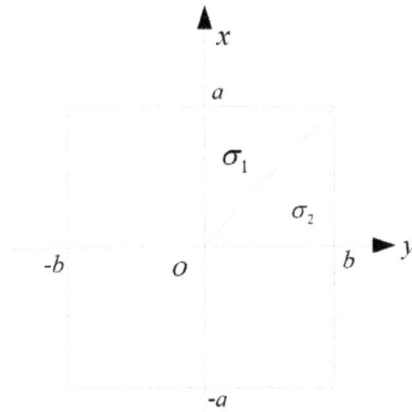

Figure 1. Integrals in the rectangular area.

As shown in **Figure 1**, let the innermost area be the rectangular $\sigma \in [-a < x < a, -b < y < b]$ $(a > 0, b > 0)$ due to the convergence of meridian, and the gravity anomaly is expressed as double quadratic polynomial:

$$\Delta g(x, y) = \sum_{i=0}^{2} \left(\frac{x}{a}\right)^i \sum_{j=0}^{2} \alpha_{ij} \left(\frac{y}{b}\right)^j \tag{29}$$

Inserting the innermost area into Eqs. (25)–(28), and let the contributions be $\Delta\zeta$, $\Delta\xi$, $\Delta\eta$ and ΔL, one arrives at

$$\Delta\zeta = \frac{1}{2\pi\gamma} \iint_\sigma \frac{\Delta g(x, y)}{r} dxdy \tag{30}$$

$$\Delta\xi = -\frac{1}{2\pi\gamma} \iint_\sigma \frac{x\Delta g(x, y)}{r^3} dxdy \tag{31}$$

$$\Delta\eta = -\frac{1}{2\pi\gamma} \iint_\sigma \frac{y\Delta g(x, y)}{r^3} dxdy \tag{32}$$

$$\Delta L = \frac{1}{2\pi} \iint_\sigma \frac{\Delta g(x, y) - \Delta g(0, 0)}{r^3} dxdy \tag{33}$$

The following transformation is introduced for σ

$$\begin{cases} u = \dfrac{x}{a} \\ v = \dfrac{y}{b} \end{cases} \tag{34}$$

Using the properties of the integration for even/odd functions and exploiting the symmetry of the integration area, one arrives at

$$\Delta\zeta = \frac{4ab}{2\pi\gamma}\int_0^1\int_0^1 \frac{\alpha_{00} + \alpha_{20}u^2 + \alpha_{02}v^2 + \alpha_{22}u^2v^2}{\sqrt{a^2u^2 + b^2v^2}}dudv \tag{35}$$

$$\Delta\xi = \frac{-4a^2b}{2\pi\gamma}\int_0^1\int_0^1 \frac{u^2(\alpha_{10} + \alpha_{12}v^2)}{(a^2u^2 + b^2v^2)^{3/2}}dudv \tag{36}$$

$$\Delta\eta = \frac{-4ab^2}{2\pi\gamma}\int_0^1\int_0^1 \frac{v^2(\alpha_{01} + \alpha_{21}v^2)}{(a^2u^2 + b^2v^2)^{3/2}}dudv \tag{37}$$

$$\Delta L = \frac{4ab}{2\pi}\int_0^1\int_0^1 \frac{\alpha_{20}u^2 + \alpha_{02}v^2 + \alpha_{22}u^2v^2}{(a^2u^2 + b^2v^2)^{3/2}}dudv \tag{38}$$

Drawing a line from the origin to the upper right corner, it divides the upper right quadrant.

into $\sigma_1 \in [0 < u < 1, 0 < v < u]$, $\sigma_2 \in [0 < u < v, 0 < u < 1]$.

The following nonsingular integration transformation is introduced for σ_1

$$\begin{cases} u = v \\ k = \dfrac{v}{u} \end{cases} \tag{39}$$

The following nonsingular integration transformation is introduced for σ_2

$$\begin{cases} v = v \\ \lambda = \dfrac{u}{v} \end{cases} \tag{40}$$

Inserting $v = ku$ (or $u = \lambda v$) into Eqs. (35)–(37), one arrives at

$$\Delta\zeta = \frac{4ab}{2\pi\gamma}\int_0^1 \left(\alpha_{00} + \frac{1}{3}\alpha_{20} + \frac{1}{3}\alpha_{02}k^2 + \frac{1}{5}\alpha_{22}k^2\right)\frac{dk}{\sqrt{a^2 + b^2k^2}}$$
$$+ \frac{4ab}{2\pi\gamma}\int_0^1 \left(\alpha_{00} + \frac{1}{3}\alpha_{02} + \frac{1}{3}\alpha_{20}\lambda^2 + \frac{1}{5}\alpha_{22}\lambda^2\right)\frac{d\lambda}{\sqrt{b^2 + a^2\lambda^2}} \tag{41}$$

$$\Delta\xi = \frac{-4a^2b}{2\pi\gamma}\int_0^1 \left(\alpha_{10} + \frac{1}{3}\alpha_{12}k^2\right)\frac{dk}{(a^2 + b^2k^2)^{3/2}}$$
$$- \frac{4a^2b}{2\pi\gamma}\int_0^1 \left(\alpha_{10} + \frac{1}{3}\alpha_{12}\right)\frac{\lambda^2 d\lambda}{(b^2 + a^2\lambda^2)^{3/2}} \tag{42}$$

$$\Delta\eta = \frac{-4ab^2}{2\pi\gamma}\int_0^1 \left(\alpha_{01} + \frac{1}{3}\alpha_{21}\right)\frac{k^2 dk}{(a^2 + b^2k^2)^{3/2}}$$
$$- \frac{4ab^2}{2\pi\gamma}\int_0^1 \left(\alpha_{01} + \frac{1}{3}\alpha_{21}\lambda^2\right)\frac{d\lambda}{(b^2 + a^2\lambda^2)^{3/2}} \tag{43}$$

$$\Delta L = \frac{4ab}{2\pi} \int_0^1 \left(\alpha_{20} + \alpha_{02}k^2 + \frac{1}{3}\alpha_{22}k^2 \right) \frac{dk}{\left(a^2 + b^2k^2\right)^{3/2}}$$

$$+ \frac{4ab}{2\pi} \int_0^1 \left(\alpha_{00} + \alpha_{20}\lambda^2 + \frac{1}{3}\alpha_{22}\lambda^2 \right) \frac{d\lambda}{\left(b^2 + a^2\lambda^2\right)^{3/2}} \qquad (44)$$

Now we can see that the denominators are greater than zero after transformation Eq. (39) and Eq. (40), and the singularities have been eliminated. The integrals in x and y directions are converted to the integrals of the powers of k and λ. This basically changes the double integrals to single variable integrals, which could easily be calculated in Mathematica as follows:

$\Delta\zeta =$

$\texttt{FullSimplify}\Big[$

$\quad \dfrac{4\,a\,b}{2\,\pi\,\gamma}\ \texttt{Integrate}\Big[\dfrac{\alpha_{0,0} + \frac{1}{3}\alpha_{2,0} + \frac{1}{3}\alpha_{0,2}\,k^2 + \frac{1}{5}\alpha_{2,2}\,k^2}{\sqrt{a^2 + b^2\,k^2}},\ \{k,\,0,\,1\},$

$\quad \texttt{Assumptions} \rightarrow \{a > 0,\ b > 0\}\Big] +$

$\quad \dfrac{4\,a\,b}{2\,\pi\,\gamma}\ \texttt{Integrate}\Big[\dfrac{\alpha_{0,0} + \frac{1}{3}\alpha_{0,2} + \frac{1}{3}\alpha_{2,0}\,\lambda^2 + \frac{1}{5}\alpha_{2,2}\,\lambda^2}{\sqrt{b^2 + a^2\,\lambda^2}},\ \{\lambda,\,0,\,1\},$

$\quad \texttt{Assumptions} \rightarrow \{a > 0,\ b > 0\}\Big]\Big]$

$\qquad (45)$

$\dfrac{1}{15\,a^2\,b^2\,\pi\,\gamma}\ \Big(b^3\ \Big(30\,a^2\,\texttt{ArcSinh}\Big[\dfrac{a}{b}\Big]\,\alpha_{0,0} +$

$\quad 10\,a^2\,\texttt{ArcSinh}\Big[\dfrac{a}{b}\Big]\,\alpha_{0,2} + \Big(a\,\sqrt{a^2 + b^2} - b^2\,\texttt{ArcSinh}\Big[\dfrac{a}{b}\Big]\Big)\,(5\,\alpha_{2,0} + 3\,\alpha_{2,2})\Big) +$

$\quad a^3\ \Big(30\,b^2\,\texttt{ArcSinh}\Big[\dfrac{b}{a}\Big]\,\alpha_{0,0} + 5\,b\,\sqrt{a^2 + b^2}\,\alpha_{0,2} + 3\,b\,\sqrt{a^2 + b^2}\,\alpha_{2,2} +$

$\quad \texttt{ArcSinh}\Big[\dfrac{b}{a}\Big]\,\big(10\,b^2\,\alpha_{2,0} - a^2\,(5\,\alpha_{0,2} + 3\,\alpha_{2,2})\big)\Big)\Big)$

$\Delta\xi =$

$\texttt{FullSimplify}\Big[-\dfrac{4\,a^2\,b}{2\,\pi\,\gamma}\ \texttt{Integrate}\Big[\dfrac{\alpha_{1,0} + \frac{1}{3}\alpha_{1,2}\,k^2}{\left(a^2 + b^2\,k^2\right)^{3/2}},\ \{k,\,0,\,1\},\ \texttt{Assumptions} \rightarrow \{a > 0,\ b > 0\}\Big] -$

$\quad \dfrac{4\,a^2\,b}{2\,\pi\,\gamma}\ \texttt{Integrate}\Big[\dfrac{\left(\alpha_{1,0} + \frac{1}{3}\alpha_{1,2}\right)\lambda^2}{\left(b^2 + a^2\,\lambda^2\right)^{3/2}},\ \{\lambda,\,0,\,1\},\ \texttt{Assumptions} \rightarrow \{a > 0,\ b > 0\}\Big]\Big]$

$$\frac{-6\,b^3\,\texttt{ArcSinh}\left[\frac{a}{b}\right]\,\alpha_{1,0} + 2\,\left(a\,b\,\sqrt{a^2 + b^2} - b^3\,\texttt{ArcSinh}\left[\frac{a}{b}\right] - a^3\,\texttt{ArcSinh}\left[\frac{b}{a}\right]\right)\alpha_{1,2}}{3\,a\,b^2\,\pi\,\gamma}$$

$\qquad (46)$

$\Delta\eta =$

$$\text{FullSimplify}\left[-\frac{4\,a\,b^2}{2\,\pi\,\gamma}\,\text{Integrate}\left[\frac{\left(\alpha_{0,1}+\frac{1}{3}\,\alpha_{2,1}\right)k^2}{\left(a^2+b^2\,k^2\right)^{3/2}},\{k,0,1\},\text{Assumptions}\rightarrow\{a>0,b>0\}\right]-\right.$$

$$\left.\frac{4\,a\,b^2}{2\,\pi\,\gamma}\,\text{Integrate}\left[\frac{\alpha_{0,1}+\frac{1}{3}\,\alpha_{2,1}\,\lambda^2}{\left(b^2+a^2\,\lambda^2\right)^{3/2}},\{\lambda,0,1\},\text{Assumptions}\rightarrow\{a>0,b>0\}\right]\right]$$

$$\frac{-6\,a^3\,\text{ArcSinh}\left[\frac{b}{a}\right]\alpha_{0,1}+2\left|a\,b\,\sqrt{a^2+b^2}-b^3\,\text{ArcSinh}\left[\frac{a}{b}\right]-a^3\,\text{ArcSinh}\left[\frac{b}{a}\right]\right|\alpha_{2,1}}{3\,a^2\,b\,\pi\,\gamma}$$

$$(47)$$

$\Delta L =$

$$\text{FullSimplify}\left[\right.$$

$$\frac{4\,a\,b}{2\,\pi}\,\text{Integrate}\left[\frac{\alpha_{2,0}+\alpha_{0,2}\,k^2+\frac{1}{3}\,\alpha_{2,2}\,k^2}{\left(a^2+b^2\,k^2\right)^{3/2}},\{k,0,1\},\text{Assumptions}\rightarrow\{a>0,b>0\}\right]+$$

$$\frac{4\,a\,b}{2\,\pi}\,\text{Integrate}\left[\frac{\alpha_{0,2}+\alpha_{2,0}\,\lambda^2+\frac{1}{3}\,\alpha_{2,2}\,\lambda^2}{\left(b^2+a^2\,\lambda^2\right)^{3/2}},\{\lambda,0,1\},\text{Assumptions}\rightarrow\{a>0,b>0\}\right]\left.\right] \quad (48)$$

$$\frac{1}{3\,a^2\,b^2\,\pi}\,2\left|3\,a^3\,\text{ArcSinh}\left[\frac{b}{a}\right]\alpha_{0,2}+3\,b^3\,\text{ArcSinh}\left[\frac{a}{b}\right]\alpha_{2,0}+\right.$$

$$\left.\left|-a\,b\,\sqrt{a^2+b^2}+b^3\,\text{ArcSinh}\left[\frac{a}{b}\right]+a^3\,\text{ArcSinh}\left[\frac{b}{a}\right]\right|\alpha_{2,2}\right|$$

In case of a square grid with a unit length, Eqs. (45)–(48) can be simplified in Mathematica as.

$$\text{FullSimplify}[\Delta\zeta\ /.\ a\rightarrow1\ /.\ b\rightarrow1]$$

$$\frac{60\,\text{ArcSinh}[1]\,\alpha_{0,0}+5\left(\sqrt{2}+\text{ArcSinh}[1]\right)(\alpha_{0,2}+\alpha_{2,0})+6\left(\sqrt{2}-\text{ArcSinh}[1]\right)\alpha_{2,2}}{15\,\pi\,\gamma} \quad (49)$$

$$\text{FullSimplify}[\Delta\xi\ /.\ a\rightarrow1\ /.\ b\rightarrow1]$$

$$\frac{-6\,\text{ArcSinh}[1]\,\alpha_{1,0}+2\left(\sqrt{2}-2\,\text{ArcSinh}[1]\right)\alpha_{1,2}}{3\,\pi\,\gamma} \quad (50)$$

$$\text{FullSimplify}[\Delta\eta\ /.\ a\rightarrow1\ /.\ b\rightarrow1]$$

$$\frac{-6\,\text{ArcSinh}[1]\,\alpha_{0,1}+2\left(\sqrt{2}-2\,\text{ArcSinh}[1]\right)\alpha_{2,1}}{3\,\pi\,\gamma} \quad (51)$$

FullSimplify[ΔL /. a → 1 /. b → 1]

$$\frac{6\,\text{ArcSinh}[1]\,(\alpha_{0,z} + \alpha_{z,0}) - 2\sqrt{2}\,\alpha_{z,z} + 4\,\text{ArcSinh}[1]\,\alpha_{z,z}}{3\,\pi} \qquad (52)$$

One could find that it is greatly fussy to complete these integrals by hand, which could be easily realized using some commands of computer algebra system.

4. The series expansions of direct transformations between three anomalies in satellite geodesy

The determination of satellite orbit is one of the fundamental problems in satellite geodesy. A graphical representation of the Keplerian orbit is given in **Figure 2**, see [20].

Eccentric, mean and true anomalies are used to describe the movement of satellites. Their transformations are often to be dealt with in satellite ephemeris computation and orbit determination of the spacecraft. In **Figure 2**, E is the Eccentric anomaly, v is the true anomaly. In order to realize the direct transformations between these anomalies, the series expansions of their transformations are derived using the power series method with the help of computer algebra system Mathematica. Their coefficients are expressed in a power series of the orbital eccentricity e and extended up to eighth-order terms of the orbital eccentricity.

4.1. The series expansions of the direct transformation between eccentric and mean anomalies

Let the mean anomaly be M. M can be expressed by E as follows:

$$M = E - e\sin E \qquad (53)$$

Differentiating the both sides of Eq. (53) yields

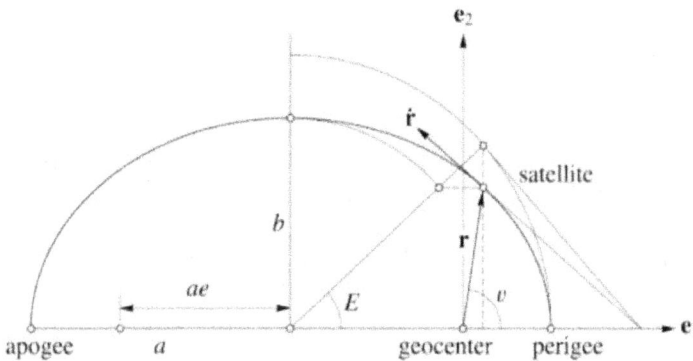

Figure 2. Keplerian orbit.

$$\frac{dE}{dM} = \frac{1}{1 - e\cos E} \tag{54}$$

To expand Eq. (54) into a power series of $\cos M$, we introduce the following new variable

$$t = \cos M \tag{55}$$

therefore

$$\frac{dM}{dt} = -\frac{1}{\sin M} \tag{56}$$

and then denote

$$f(t) = \frac{dE}{dM} = \frac{1}{1 - e\cos E} \tag{57}$$

Substituting $E_0 = \frac{\pi}{2}$ into Eq. (53) yields

$$M_0 = \frac{\pi}{2} - e \tag{58}$$

Substituting Eq. (59) into Eq. (55), one arrives at

$$t_0 = \sin e \tag{59}$$

Making use of the chain rule of implicit differentiation

$$f'(t) = \frac{df}{dE}\frac{dE}{dM}\frac{dM}{dt}$$

$$f''(t) = \frac{df'}{dE}\frac{dE}{dM}\frac{dM}{dt} + \frac{df'}{dM}\frac{dM}{dt}$$

$$\cdots$$

It is easy to expand Eq. (58) into a power series of t_0

$$f(t) = \frac{dE}{dM} = f(t_0) + f'(t_0)(t - t_0) + \frac{1}{2!}f''(t_0)(t - t_0)^2 + \frac{1}{3!}f'''(t_0)(t - t_0)^3 + \cdots \tag{60}$$

One can imagine that these procedures are too complicated to be realized by hand, but will become much easier and be significantly simplified by means of Mathematica. Omitting the detailed procedure in Mathematica, one arrives at

$$\frac{dE}{dM} = 1 + b_1(\cos M - \sin e) + \frac{b_2}{2!}(\cos M - \sin e)^2 + \frac{b_3}{3!}(\cos M - \sin e)^3 + \frac{b_4}{4!}(\cos M - \sin e)^4$$

$$+ \frac{b_5}{5!}(\cos M - \sin e)^5 + \frac{b_6}{6!}(\cos M - \sin e)^6 + \frac{b_7}{7!}(\cos M - \sin e)^7 + \frac{b_8}{8!}(\cos M - \sin e)^8 \tag{61}$$

where

$$
\begin{cases}
b_1 = e + \frac{1}{2}e^3 + \frac{1}{24}e + 5\frac{61}{720}e^7 \\[2mm]
b_2 = 4e^2 + \frac{13}{3}e^4 + \frac{47}{15}e^6 + \frac{121}{63}e^8 \\[2mm]
b_3 = 27e^3 + \frac{91}{2}e^5 + \frac{1127}{24}e^7 \\[2mm]
b_4 = 256e^4 + \frac{2937}{5}e^6 + \frac{82771}{105}e^8 \\[2mm]
b_5 = 3125e^5 + \frac{18173}{2}e^7 \\[2mm]
b_6 = 46656e^6 + \frac{1150593}{7}e^8 \\[2mm]
b_7 = 823543e^7 \\[2mm]
b_8 = 16777216e^8
\end{cases}
\tag{62}
$$

Integrating at the both sides of Eq. (62) gives the series expansion

$$
\begin{aligned}
E = M &+ \alpha_1 \sin M + \alpha_2 \sin 2M + \alpha_3 \sin 3M + \alpha_4 \sin 4M + \alpha_5 \sin 5M \\
&+ \alpha_6 \sin 6M + \alpha_7 \sin 7M + \alpha_8 \sin 8M
\end{aligned}
\tag{63}
$$

where

$$
\begin{cases}
\alpha_1 = e - \frac{1}{8}e^3 + \frac{1}{192}e^5 - \frac{1}{9216}e^7 \\[2mm]
\alpha_2 = \frac{1}{2}e^2 - \frac{1}{6}e^4 + \frac{1}{48}e^6 - \frac{1}{720}e^8 \\[2mm]
\alpha_3 = \frac{3}{8}e^3 - \frac{27}{128}e^5 + \frac{243}{5120}e^7 \\[2mm]
\alpha_4 = \frac{1}{3}e^4 - \frac{4}{15}e^6 + \frac{4}{45}e^8 \\[2mm]
\alpha_5 = \frac{125}{384}e^5 - \frac{3125}{9216}e^7 \\[2mm]
\alpha_6 = \frac{27}{80}e^6 - \frac{243}{560}e^8 \\[2mm]
\alpha_7 = \frac{16807}{46080}e^7 \\[2mm]
\alpha_8 = \frac{128}{315}e^8
\end{cases}
\tag{64}
$$

4.2. The series expansions of the direct transformation between eccentric and true anomalies

The true anomaly v can be expressed by E as follows:

$$\tan \frac{v}{2} = \sqrt{\frac{1+e}{1-e}} \tan \frac{E}{2} \tag{65}$$

Therefore, it holds

$$v = 2\arctan\left(\sqrt{\frac{1+e}{1-e}} \tan \frac{E}{2}\right) \tag{66}$$

One could expand v as a power series of the eccentricity at $e = 0$ in order to obtain the direct series expansion of the transformation from E to v. Omitting the detailed procedure in Mathematica, one arrives at

$$\begin{aligned} v = E + \beta_1 \sin E + \beta_2 \sin 2E + \beta_3 \sin 3E + \beta_4 \sin 4E + \beta_5 \sin 5E \\ + \beta_6 \sin 6E + \beta_7 \sin 7E + \beta_8 \sin 8E \end{aligned} \tag{67}$$

where

$$\begin{cases} \beta_1 = e + \dfrac{1}{4}e^3 + \dfrac{1}{8}e^5 + \dfrac{5}{64}e^7 \\[2mm] \beta_2 = \dfrac{1}{4}e^2 + \dfrac{1}{8}e^4 + \dfrac{5}{64}e^6 + \dfrac{7}{128}e^8 \\[2mm] \beta_3 = \dfrac{1}{12}e^3 + \dfrac{1}{16}e^5 + \dfrac{3}{64}e^7 \\[2mm] \beta_4 = \dfrac{1}{32}e^4 + \dfrac{1}{32}e^6 + \dfrac{7}{256}e^8 \\[2mm] \beta_5 = \dfrac{1}{80}e^5 + \dfrac{1}{64}e^7 \\[2mm] \beta_6 = \dfrac{1}{192}e^6 + \dfrac{1}{128}e^8 \\[2mm] \beta_7 = \dfrac{1}{448}e^7 \\[2mm] \beta_8 = \dfrac{1}{1024}e^8 \end{cases} \tag{68}$$

From Eq. (67), one knows

$$E = 2\arctan\left(\sqrt{\frac{1-e}{1+e}} \tan \frac{v}{2}\right) \tag{69}$$

Expanding E as a power series of the eccentricity at $e = 0$ by means of Mathematica yields the direct series expansion of the transformation from v to E

$$E = v + \gamma_1 \sin v + \gamma_2 \sin 2v + \gamma_3 \sin 3v + \gamma_4 \sin 4v + \gamma_5 \sin 5v$$
$$+ \gamma_6 \sin 6v + \gamma_7 \sin 7v + \gamma_8 \sin 8v$$
(70)

where

$$
\begin{cases}
\gamma_1 = -e - \dfrac{1}{4}e^3 - \dfrac{1}{8}e^5 - \dfrac{5}{64}e^7 \\[2mm]
\gamma_2 = \dfrac{1}{4}e^2 + \dfrac{1}{8}e^4 + \dfrac{5}{64}e^6 + \dfrac{7}{128}e^8 \\[2mm]
\gamma_3 = -\dfrac{1}{12}e^3 - \dfrac{1}{16}e^5 - \dfrac{3}{64}e^7 \\[2mm]
\gamma_4 = \dfrac{1}{32}e^4 + \dfrac{1}{32}e^6 + \dfrac{7}{256}e^8 \\[2mm]
\gamma_5 = -\dfrac{1}{80}e^5 - \dfrac{1}{64}e^7 \\[2mm]
\gamma_6 = \dfrac{1}{192}e^6 + \dfrac{1}{128}e^8 \\[2mm]
\gamma_7 = -\dfrac{1}{448}e^7 \\[2mm]
\gamma_8 = \dfrac{1}{1024}e^8
\end{cases}
$$
(71)

4.3. The series expansions of the direct transformation between mean and true anomalies

The whole formulae for the transformation from M to v are as follows

$$
\begin{cases}
E = M + \alpha_1 \sin M + \alpha_2 \sin 2M + \alpha_3 \sin 3M + \alpha_4 \sin 4M + \alpha_5 \sin 5M \\[2mm]
\quad + \alpha_6 \sin 6M + \alpha_7 \sin 7M + \alpha_8 \sin 8M \\[2mm]
v = E + \beta_1 \sin E + \beta_2 \sin 2E + \beta_3 \sin 3E + \beta_4 \sin 4E + \beta_5 \sin 5E \\[2mm]
\quad + \beta_6 \sin 6E + \beta_7 \sin 7E + \beta_8 \sin 8E
\end{cases}
$$
(72)

Since the coefficients α_i, β_i ($i = 1, 2, \cdots 8$) are expressed in a power series of the eccentricity, one could expand v as a power series of the eccentricity at $e = 0$ in order to obtain the direct expansion of the transformation from M to v. Omitting the main operations by means of Mathematica, one arrives at the direct expansion of the transformation from M to v

$$v = M + \delta_1 \sin M + \delta_2 \sin 2M + \delta_3 \sin 3M + \delta_4 \sin 4M + \delta_5 \sin 5M$$
$$+ \delta_6 \sin 6M + \delta_7 \sin 7M + \delta_8 \sin 8M$$
(73)

where

$$
\left\{
\begin{aligned}
\delta_1 &= 2e - \frac{1}{4}e^3 + \frac{5}{96}e^5 + \frac{107}{4608}e^7 \\
\delta_2 &= \frac{5}{4}e^2 - \frac{11}{24}e^4 + \frac{17}{192}e^6 + \frac{43}{5760}e^8 \\
\delta_3 &= \frac{13}{12}e^3 - \frac{43}{64}e^5 + \frac{95}{512}e^7 \\
\delta_4 &= \frac{103}{96}e^4 - \frac{451}{480}e^6 + \frac{4123}{11520}e^8 \\
\delta_5 &= \frac{1097}{960}e^5 - \frac{5957}{4608}e^7 \\
\delta_6 &= \frac{1223}{960}e^6 - \frac{7913}{4480}e^8 \\
\delta_7 &= \frac{47273}{32256}e^7 \\
\delta_8 &= \frac{556403}{322560}e^8
\end{aligned}
\right.
\tag{74}
$$

The whole formulae for the transformation from v to M are as follows

$$
\left\{
\begin{aligned}
E &= v + \gamma_1 \sin v + \gamma_2 \sin 2v + \gamma_3 \sin 3v + \gamma_4 \sin 4v + \gamma_5 \sin 5v \\
&\quad + \gamma_6 \sin 6v + \gamma_7 \sin 7v + \gamma_8 \sin 8v \\
M &= E - e \sin E
\end{aligned}
\right.
\tag{75}
$$

Expanding M as a power series of the eccentricity at $e = 0$ by means of Mathematica yields the direct series expansion of the transformation from v to M

$$
\begin{aligned}
M &= v + \varepsilon_1 \sin v + \varepsilon_2 \sin 2v + \varepsilon_3 \sin 3v + \varepsilon_4 \sin 4v \\
&\quad + \varepsilon_5 \sin 5v + \varepsilon_6 \sin 6v + \varepsilon_7 \sin 7v + \varepsilon_8 \sin 8v
\end{aligned}
\tag{76}
$$

where

$$
\left\{
\begin{aligned}
\varepsilon_1 &= -2e \\
\varepsilon_2 &= \frac{3}{4}e^2 + \frac{1}{8}e^4 + \frac{3}{64}e^6 + \frac{3}{128}e^8 \\
\varepsilon_3 &= -\frac{1}{3}e^3 - \frac{1}{8}e^5 - \frac{7}{16}e^7 \\
\varepsilon_4 &= \frac{5}{32}e^4 + \frac{3}{32}e^6 + \frac{15}{256}e^6 \\
\varepsilon_5 &= -\frac{3}{40}e^5 - \frac{1}{16}e^7 \\
\varepsilon_6 &= \frac{7}{192}e^6 + \frac{5}{128}e^6 \\
\varepsilon_7 &= -\frac{1}{56}e^7 \\
\varepsilon_8 &= \frac{9}{1024}e^8
\end{aligned}
\right.
\tag{77}
$$

$E_0/(°)$	20	40	60	80
$\Delta E_1/(\prime\prime)$	8.2×10^{-8}	-1.7×10^{-7}	1.6×10^{-7}	2.2×10^{-8}
$\Delta v_1/(\prime\prime)$	3.5×10^{-7}	-7.4×10^{-7}	6.8×10^{-7}	9.9×10^{-8}
$\Delta M_1/(\prime\prime)$	3.5×10^{-8}	-5.2×10^{-9}	-7.1×10^{-9}	-3.7×10^{-9}
$\Delta E_2/(\prime\prime)$	2.8×10^{-8}	2.1×10^{-8}	1.4×10^{-8}	1.2×10^{-8}
$\Delta v_2/(\prime\prime)$	-2.9×10^{-8}	-2.4×10^{-8}	-1.5×10^{-8}	-1.3×10^{-8}

Table 2. Errors of the derived series expansions.

4.4. The accuracy of the derived series expansions

In order to validate the exactness of the derived series expansions, one has examined their accuracies when the orbital eccentricity e is respectively equal to 0.01, 0.05, 0.1 and 0.2.

One makes use of Eq. (53) and Eq. (67) to obtain the theoretical value M_0 and v_0 at given geodetic latitude E_0. Substituting M_0 into Eq. (64) and Eq. (74), one arrives at the computation value E_1 and v_1. Substituting v_0 into Eq. (71) and Eq. (77), one arrives at the computation value E_2 and M_1. Substituting E_0 into Eq. (68), one arrives at the computation value v_2. The differences between the computation and theoretical values indicate the accuracies of the derived series expansions, which are denoted as ΔE_1, $\Delta v_1/(\prime\prime)$, $\Delta M_1/(\prime\prime)$, $\Delta E_2/(\prime\prime)$, $\Delta v_2/(\prime\prime)$. Due to limited space, these errors when e is equal to 0.05 are only listed in **Table 2**.

From **Table 2**, one could find that the accuracy of derived series expansions is higher than $10^{-5\prime\prime}$, which could satisfy practical application. Other numerical examples indicate that when the orbital eccentricity e is respectively equal to 0.01, 0.01 and 0.2, the accuracy of derived series expansions is correspondingly higher than $10^{-10\prime\prime}$, $10^{-3\prime\prime}$ and $0.1^{\prime\prime}$.

5. Conclusions

Some typical mathematical problems in geodesy are solved by means of computer algebra analysis method and computer algebra system Mathematica. The main contents and research results presented in this chapter are as follows:

1. The forward and inverse expansions of the meridian arc often used in geometric geodesy are derived. Their coefficients are expressed in a power series of the first eccentricity of the reference ellipsoid and extended up to its tenth-order terms.

2. The singularity existing in the integral expressions of height anomaly, deflections of the vertical and gravity gradient is eliminated using the nonsingular integration transformations, and the nonsingular expressions are systematically derived.

3. The series expansions of direct transformations between three anomalies in satellite geodesy are derived using the power series method. Their coefficients are expressed in a power series of the orbital eccentricity e and extended up to eighth-order terms of the orbital eccentricity.

Acknowledgements

This work was financially supported by National Natural Science Foundation of China (Nos. 41631072, 41771487, 41571441).

Conflict of interest

There are no conflicts of interest.

Thanks

We would like to express our great appreciation to the editor and reviewers. Thanks very much for the kind work and consideration of Ms. Kristina Jurdana on the publication of this chapter.

Author details

Hou-pu Li* and Shao-feng Bian

*Address all correspondence to: lihoupu1985@126.com

Department of Navigation, Naval University of Engineering, Wuhan, China

References

[1] Adam OS. Latitude developments connected with geodesy and cartography with tables, including a table for Lambert equal-area meridional projection. U.S. Coast and Geodetic Survey, Spec. Pub. No. 67; 1921

[2] Deakin RE, Hunter MN. Geometric Geodesy Part A, RMIT University; 2013

[3] Hofmann-Wellenhof B, Moritz H. Physical Geodesy. Wien: Springer-Verlag; 2005

[4] Seeber G. Satellite geodesy. Berlin, New York: Walter de Gruyter; 2003

[5] Jie X. Ellipsoidal Geodesy. Beijing: PLA Press; 1988. (in Chinese)

[6] Yang Q. The Theories and Methods of Map Projection. Beijing: PLA Press; 1989. (in Chinese)

[7] Yang QH, Snyder JP, Tobler WR. Map Projection Transformation: Principles and Applications. London: Taylor and Francis; 2000

[8] Huatong Z. Geodetic Computing for Ellipsoidal Geodesy. Surveying and Mapping Press; 1993. (in Chinese)

[9] Lv ZP, Qy Y, Qiao SB. Geodesy Introduction to Geodetic Datum and Geodetic Systems. Berlin: Springer; 2014

[10] Awange JL, Grafarend EW. Solving Algebraic Computational Problems in Geodesy and Geoinformatics. Berlin: Springer; 2005

[11] Awange JL, Grafarend EW, Paláncz B, Zaletnyik P. Algebraic Geodesy and Geoinformatics. Berlin: Springer; 2010

[12] Shaofeng B, Houpu L. Mathematical analysis in cartography by means of computer algebra system. In: Bateira C, editor. Cartography—A Tool for Spatial Analysis. InTech; 2012. pp. 1-24

[13] Bian SF, Chen YB. Solving an inverse problem of a meridian arc in terms of computer algebra system. Journal of Surveying Engineering. 2006;**132**(1):153-155. DOI: 10.1061/(ASCE)0733-9453(2006)132:1(7)

[14] Shaofeng B, Jiangning X. The Computer Algebra System and Mathematical Analysis in Geodesy. Beijing: National Defense Industry Press; 2004. (in Chinese)

[15] Houpu L, Shaofeng B, Zhongbin. Precise Analysis of Geographic Cooridinate System by Means of Computer Algebra, Beijing. National Defense Industry Press; 2015. (in Chinese)

[16] Hwang C, Hsu HY, Jang RJ. Global mean sea surface and marine gravity anomaly from multi-satellite altimetry: Applications of deflection-geoid and inverse Vening Meinesz formulae. Journal of Geodesy. 2002;**76**:407-418. DOI: 10.1007/s00190-002-0265-6

[17] Sunkel H. Die Darstellung Geodtischer Integral formeln durch Bi-Kubische Spline Funktionen. Folge 28. Graz; 1977

[18] Wang Y. Problem der Glattung bei den Integraloperation der Physikallischen Geodasie. Folge 55. Graz; 1987

[19] Shaofeng B, Huangqi S. The expression of common singular integrals in physical geodesy. Manuscripta Geodaetica. 1994;**19**:62-69

[20] Hofmann-Wellenhof B, Lichtenegger H, Wasle E. GNSS-Global Navigation Satellite Systems GPS, GLONASS, Galileo, and More. New York: Springer-Verlag Wien; 2008